SpringerBriefs in Applied Sciences and Technology

Safety Management

Series Editors

Eric Marsden, FonCSI, Toulouse, France

Caroline Kamaté, FonCSI, Toulouse, France

François Daniellou, FonCSI, Toulouse, France

The SpringerBriefs in Safety Management present cutting-edge research results on the management of technological risks and decision-making in high-stakes settings.

Decision-making in high-hazard environments is often affected by uncertainty and ambiguity; it is characterized by trade-offs between multiple, competing objectives. Managers and regulators need conceptual tools to help them develop risk management strategies, establish appropriate compromises and justify their decisions in such ambiguous settings. This series weaves together insights from multiple scientific disciplines that shed light on these problems, including organization studies, psychology, sociology, economics, law and engineering. It explores novel topics related to safety management, anticipating operational challenges in high-hazard industries and the societal concerns associated with these activities.

These publications are by and for academics and practitioners (industry, regulators) in safety management and risk research. Relevant industry sectors include nuclear, offshore oil and gas, chemicals processing, aviation, railways, construction and healthcare. Some emphasis is placed on explaining concepts to a non-specialized audience, and the shorter format ensures a concentrated approach to the topics treated.

The SpringerBriefs in Safety Management series is coordinated by the Foundation for an Industrial Safety Culture (FonCSI), a public-interest research foundation based in Toulouse, France. The FonCSI funds research on industrial safety and the management of technological risks, identifies and highlights new ideas and innovative practices, and disseminates research results to all interested parties.

For more information: https://www.foncsi.org/.

More information about this subseries at http://www.springer.com/series/15119

Corinne Bieder · Kenneth Pettersen Gould
Editors

The Coupling of Safety and Security

Exploring Interrelations in Theory and Practice

Editors
Corinne Bieder
Safety—Security research program
ENAC, University of Toulouse
Toulouse, France

Kenneth Pettersen Gould
Department of Safety,
Economics and Planning
University of Stavanger
Stavanger, Norway

ISSN 2191-530X   ISSN 2191-5318   (electronic)
SpringerBriefs in Applied Sciences and Technology
ISSN 2520-8004   ISSN 2520-8012   (electronic)
SpringerBriefs in Safety Management
ISBN 978-3-030-47228-3   ISBN 978-3-030-47229-0   (eBook)
https://doi.org/10.1007/978-3-030-47229-0

This Springer imprint is published by the registered company Springer Nature Switzerland AG
The registered company address is: Gewerbestrasse 11, 6330 Cham, Switzerland

Preface

Safety has long been a major concern for hazardous industries. With the increase in security threats over the past two decades, safety and security have come together and now coexist as strategies and management practices. However, they often do so, without thoughtful reflection about their interrelations and the kind of implications this may have.

Investigating this area from diverse perspectives and identifying the synergies and tensions between safety and security was at the core of a 3-day workshop organized by the NeTWork[1] think tank and gathering researchers from different disciplines and countries. This workshop was held in the inspiring Abbaye of Royaumont, near Paris, in June 2018. Engaging in this exchange brought enlightening insights to the complex interrelations between safety and security but also to the associated research and management challenges.

The co-editors, Corinne Bieder and Kenneth Pettersen Gould, are deeply grateful to the FonCSI (Foundation for an Industrial Safety Culture)[2] for the support and funding of this research initiative.

Toulouse, France
Stavanger, Norway

Corinne Bieder
Kenneth Pettersen Gould

---

[1]NeTWork: http://www.network-network.org/.
[2]FonCSI: https://www.foncsi.org/.

Contents

Chapter 1
Safety and Security: The Challenges of Bringing Them Together

Kenneth Pettersen Gould and Corinne Bieder

Abstract This chapter looks back at how safety and security have developed in hazardous technologies and activities, explaining what has become an intersection between the two in both strategies and management practices. We argue for the connection to be made between social expectations of safe and secure societies and the limits to management and technical performance. In the first part of the chapter, conceptual similarities and differences are addressed and we distinguish three scientific and contextual vantage points for addressing how safety and security are converging: the conceptual approach, the technical and methodological approach, and the management and practice approach. We then go on to show that, as professional areas, safety and security have developed in different ways and supported by quite separate scientific and technological fields. Finally, we present the organization of the book.

Keywords Safety · Security · Science · Management · Societal safety · Societal security

The exploratory title of this book aims to encourage the reader to think about the development of safety and security in combination and with renewed perspectives. A key background for this bringing together of concepts is the general trend in society that the safer and more secure our organizations and institutions become, the more of it we demand from them. While many of the biggest threats to health and safety at work have been reduced, at least in Europe and North America, industrial safety has become broadened through increased emphasis on modern societies' production of new systemic risks and the ideas that vulnerabilities are affected by global events [2, 3, 6, 22].

K. Pettersen Gould (✉)
University of Stavanger, Stavanger, Norway
e-mail: kenneth.a.pettersen@uis.no

C. Bieder
ENAC (French Civil Aviation University), University of Toulouse, Toulouse, France
e-mail: corinne.bieder@enac.fr

© The Author(s) 2020
C. Bieder and K. Pettersen Gould (eds.), *The Coupling of Safety and Security*, SpringerBriefs in Safety Management,
https://doi.org/10.1007/978-3-030-47229-0_1

1.1 How Do Safety and Security Come Together?

Safety has long been a major concern for organizations, especially with the advent of hazardous technologies and activities. Within sectors such as energy, chemical, transportation, water, and health, safety is a core concept in policy, regulation, and management. Consequently, there are well-established institutional/management strategies, collaborations, and practices associated with preventing incidents and accidents. Maintaining the efficacy of these approaches is viewed as important for protecting hazardous technologies, as they are based on previous incidents and include the dynamic yet fragile organizational web of safety defenses [24]. From the 1980s, supported by an increased understanding of how and why accidents happen, increased attention was paid to how accidents and disasters are caused by societal developments [6]. Research demonstrated how hazards relate to changing organizational characteristics [12, 21], and the argument that major accidents are inevitable in certain high-hazard systems became influential and spurred interest in the limits to safety and possibilities of organizational competence [12].

Security was up until the end of the Cold War strongly connected to state security and the protection against threats from foreign states. For civilian industries, security in this respect became an issue to the extent that organizations contributed to a state's military defense capabilities [6]. However, when the Cold War ended in the late 1980s, the political focus shifted to peace and international human rights as well as an increased consciousness about societies' own vulnerabilities to malicious acts such as sabotage and terrorism. Until the catastrophic attacks in New York on the 11th of September 2001, security threats were a much smaller part of the overall regulatory and management scope compared to other hazards considered[1] (i.e., major accidents and disasters). However, now over 15 years later, we have become far more familiar with facing malicious attacks that may involve suicide operations. Partly because of this change in the type of threat, the public feels a form of free-floating dread which is amplified by terrorist attacks [10]. New public policy notions have been introduced and different management and/or organizational perspectives are established for a secure society. Better preparation, emphasizing prevention in particular, has been called for by the public with new demands and accountabilities being developed [10]. In the U.S., the Transportation Security Administration (TSA), now part of the Homeland Security Department, was created following the creation and approval of the Aviation and Transportation Security Act in November 2001 (9/11 Commission report, 2004). As illustrated later in this chapter, similar developments have been seen in European countries.

The increasing emphasis on security and associated security risk reduction measures leads to an obvious intersection between safety and security management in hazardous industries. Leaders and analysts have had to understand and include a new category of threats. New forms of cooperation and domains of operations have also developed, that were not key premises in existing strategies and practices prior to

---

[1]There are also further categories of organizational practice such as Workplace Health and Safety and cybersecurity with similar developments in hazardous industries.

9/11. In addition, doubt has been cast on the efficacy of a good deal of the existing approaches to protecting hazardous technologies [10]. The interactions between safety and security have emerged as not all obvious, especially in normal situations. Subtle mutual influences do occur. As illustrated by Pettersen & Bjørnskau [13], safety and security practices may in some cases conflict with one another. Many organizations (industries, institutions) are hesitating whether they should have two separate entities for managing safety and security or merge the two.

For both safety and security, hazards and threats are today defined more and more as systemic risks and products of modern society. Local vulnerabilities are increasingly being understood as influenced by global events and processes [6], such as within digitalization [9, 19]. These developments coincide with a similar transformation of both safety and security policy, toward broader fields and shared responsibilities focusing on societal, civil, homeland, and human issues. These changes must also be viewed in combination with a growth in risk management as solutions to policy requirements [6, 17, 20], developments that are connected to a wider pattern of neo-liberal influence [14] characterized by extensive deregulation, privatization, and outsourcing. Where the gray area between security and safety previously could be narrowed to the problem of defining the difference between an accident and a criminal act, safety and security can no longer (if they ever did) ignore each other in either concepts, policy or management practice.

1.2   Safety and Security

Although there may be little difference between feeling secure and feeling safe [1], if we admit that the concepts of safety and security are not fully analogous, providing clear definitions of the concepts remains a challenge [4]. Not only is there a single word for safety and security in many languages (unlike in English), but also the many definitions from academics on the one hand and the colloquial use of the terms on the other hand convey ambiguities [4].

The definitions provided by academics mainly refer to two types of distinctions between safety and security: one related to the intentionality, safety focusing on hazards and non-intentional or accidental risks as opposed to security that focuses on malicious threats and intentional risks [1, 20]. The other one builds on the differences of origins—consequences, safety being the ability of the system not to harm the environment whereas security is the ability of the environment not to harm the system [4, 15]. Yet, further refinements are proposed by some authors combining these two axes of distinction between safety and security, especially to account for differences in the use of terms in different domains and to enrich the system—environment axis by considering the ability of a system not to harm itself [16].

Despite efforts at refining the distinction between safety and security, a returning question is whether to distinguish the two or to best manage dangers overall whether they make us feel unsafe or insecure [23]. A central concept for how to achieve both safety and security is risk management (Blokland & Reniers, Bieder

& Pettersen Gould, both this volume). However, there is much confusion as to what to expect of risk analysis [18], how it can be carried out, and if it is the same for safety and security [8]. The conceptual differences between safety and security have in many contexts become further extended by science and technology, for example, in airport operations [5, 6]. In daily operations, security screeners and safety personnel have different training, use different technologies, and operate in completely different ways. The different regulatory frameworks and the nature of some of the contracts in place at airports [5] further strengthen this divide. Still, such behavior by individual workers or organizations to protect against or mitigate threats and hazards requires decisions without clarification whether it is a matter of safety or security. Many such decisions involve "ordinary workers", managers, as well as HSE professionals, security officers, and other professionals. Blurred as these distinctions are, the contributions included in this book are a sample of how safety and security are converging based on different scientific positions and contextual vantage points: conceptual (Blokland & Reniers; Jore), technical and methodological (Leveson; Wipf; Bongiovanni), and management and practice (Brooks & Coole; La Porte; Boustras; Schulman). The chapters show that doing both safety and security are quite generic features of organizations and for many integral to their existence. However, distinguished as professional areas, safety and security have developed in different ways and supported by quite separate scientific and technological fields. Still, while some areas of professional safety and security practice are supported by highly specialized and rigorous knowledge, others are routinized by convention, rule, or law [18]. This requires, in addition to technical knowledge and methods, learning by empirical study of the organizations and systems in which safety and security develop and interact (see La Porte, this volume).

1.3 Safety, Security, Science, and Public Policy

Both safety and security can claim to be relatively young as scientific communities [20, 11]. Safety science is customarily described as research for increased protection, preventing danger or risk of injury. However, the production of safety knowledge has proven to be diverse, with variations depending on context and with a mix of approaches from different disciplines [11]. Safety managers are also a diverse community that continues to grow, but some boundaries have developed, for example, within Occupational Health and Safety (OHS) [7]. Despite this, technical demands for safety vary a great deal, depending on the type of hazards that are at issue, and many requirements are legal or economic and originate more from policy, rather than from science. More specifically, they originate from the institutional and organizational goals in the safety strategies and climate of supranational regulatory agency, national or local government, or corporation. As for security science, it is as diverse as safety, equally multidisciplined, and with an even less defined knowledge and skill structure [20]. However, as for safety, security can be defined more clearly when you connect it to a specific context and supporting concepts, theories, and models that can be identified (Ibid.).

The chapters that follow focus on safety and security as distinct practices, such as Brooks and Coole investigate in Chap. 7, drawing attention to the role of practices and professions, but we also learn a great deal about safety science and security science. For example, Sissel Jore notes that an accident investigation report used security culture as one important explanatory factor behind the outcome of a terrorist attack and that many Norwegian petroleum companies apply security culture as a means of security improvement. While clearly having its counterpart in safety culture and in theory possible to define and investigate, the concept is applied with little technical support and analysis. Distinctions between security and safety as well as between scientific approaches and management thus become blurred both in theory and in practice.

1.4 Safety, Security, and Social Expectations

That there are limits to bureaucratic and technical performance in the search for safety and security is undeniable. Both accidents and malicious attacks will happen and uncertainties will continue to abound [18]. As Schulman states in Chap. 9, there are always more ways that a complex system can fail than there are for it to operate correctly as designed. And hostile strategy can add additional possibilities for disaster because of the treatment of vulnerabilities as strategic targets. Yet one must ask, what is safe and secure enough? Also, as safety and security prompt new demands, even for stronger integration, what are the implications for the people in organizations and institutions managing technologies that continue to grow in scale and complexity? Who will be reinforced and who will experience increased tension and conflict? See La Porte (this volume).

As previously stated, many organizations (industries, institutions) are hesitating whether they should have two separate entities to manage safety and security or merge their treatment. Security is being added to the scope of some safety authorities (for example, in aviation with EASA and the French Civil Aviation Authority), but with very limited inputs from research as to how to deal conceptually and in practice with this extended scope. A further potential issue is around transparency and sharing of data and experience. An illustration is the publication of research, where security research may demand confidentiality about results, whereas safety management practice and safety research aim for maximum openness.

So far, most research and literature on the relationship between safety and security has focused on engineering aspects like design and risk analysis methods [15], as well as some work on conceptual issues [4]. But despite the number of years where safety and security have coexisted as approaches, there seems to be limited research on how safety and security are managed in practice at all levels. Importantly, a few field studies confirm a tension between safety and security when it comes to daily activities [5, 13] and thus there is a need to further investigate the interactions between the two aspects within hazardous technologies and activities.

1.5 Organization of the Book

The chapters of this book are not put into any divided sections, but the flow of chapters follows roughly their focus from conceptual, technical, and methodological themes toward issues of empirical research, management, and practice. It is instructive to note, however, that there is a good deal of overlap between chapters. The final chapter summarizes some key challenges and problems from looking across contributions and discusses some key issues for an interrelated research agenda for safety and security. In chapter two, Blokland and Reniers take a risk perspective and focus on what links and differentiates safety and security in situations where there is uncertainty related to effects on individual, organizational, or societal objectives. For risk analysis purposes, the chapter largely outlines safety and security in the same way but also argues for some differences between security and safety related to effects, objectives, and uncertainty. Leveson, in chapter three, presents how system safety engineering methods can be developed to include both safety and security scenarios. The approach taken acknowledges that system design errors cannot be eliminated prior to use and that the complexity of many systems requires new and more inclusive models of causality. The chapter illustrates how engineering tools based on system theory can be applied to handle safety and security in an integrated manner. Based on an empirical case from light helicopter operations, Wipf uses a game theory approach in chapter four to assess safety and security issues in combination. The chapter illustrates the commonalities and differences between assessment techniques. Acknowledging the turn of security science toward softer measures, Jore argues for security culture as a promising concept for organizations, as it can make security a priority and shared responsibility, and compares it in chapter five to the more widely applied concept safety culture. The adequacy of the concept is discussed based on its use in an investigation report from a terrorist attack on an internationally run Algerian oil facility and the discussion is structured by using criteria for conceptual adequacy. Chapter 6 is methods oriented and takes an end user perspective on safety and security. Focusing on an airport security environment and security screening in particular, Bongiovanni shows the potential benefits of looking beyond legal and managerial perspectives that seem to dominate both safety and security management. He argues that this can help organizations to use fewer resources on the "eternal killjoys" of loss prevention and increase value for users. Chapter 7 explains how safety and security, though sharing an overarching drive for social welfare, are diverging as distinct professions. Brooks and Coole explain, considering security within the context of corporate security and safety within the context of occupational health and safety, that when considered within their occupations and supporting bodies of professional knowledge, there are limited synergies. "What organizational design and operational puzzles arise when 'safety in operation', and then 'security from external threat' are demanded from organizations and public institutions as their core technologies grow in scale and complexity" asks LaPorte in Chap. 8. Building on experience from a field study of large-scale technical organizations, the chapter formulates questions that emerge when safety and security become mixed operational challenges and sketches out a guide toward further empirical research. The chapter also addresses strategic implications for senior leadership confronted by both external threats and the increasing operating social complexities of organizations operating hazardous systems. Based on previous research on high-reliability management, Schulman focuses in chapter nine on the management challenge of safety and security convergence. He discusses how high reliability may function as a common framework for safety and security, as well as challenges of bringing safety and security under a larger management framework. And in Chap. 10, Boustras explores safety and security from the perspective of the workplace, arguing for how emerging risks and new drivers are motivating new focus areas in the interface of safety and security. As job-related consequences and the direct economic impact for organizations are less apparent, state authorities and regulatory pressure become more of the backbone but with increasing demands on the workplace.

References

1. B. Ale, *Risk: an Introduction: the Concepts of Risk, Danger and Chance*. Routledge (2009)
2. U. Beck, *World at Risk*. Polity (2009)
3. U. Beck, From industrial society to the risk society: questions of survival, social structure and ecological enlightenment. Theory Culture Soc. **9**(1), 97–123 (1992)
4. Boholm et al., The concepts of risk, safety, and security: applications in everyday language, Risk Anal. **36**(2) (2016)
5. I. Bongiovanni, Assessing vulnerability to safety and security disruptions in Australian airports (Doctoral dissertation, Queensland University of Technology) (2016)
6. O.A.H. Engen, B.I. Kruke, P.H. Lindøe, K.H. Olsen, O.E. Olsen, K.A. Pettersen, Perspektiver på samfunnssikkerhet. Perspectives on societal security, Cappelen Damm (2016)
7. A. Hale, From national to European frameworks for understanding the role of occupational health and safety (OHS) specialists. Saf. Sci. **115**, 435–445 (2019)
8. S.H. Jore, The conceptual and scientific demarcation of security in contrast to safety. Eur. J. Secur. Re. **4**(1), 157–174 (2019)
9. S. Kriaa, L. Pietre-Cambacedes, M. Bouissou, Y. Halgand, A survey of approaches combining safety and security for industrial control systems. Reliab. Eng. Syst. Saf. **139**, 156–178 (2015)
10. T.R. LaPorte, Challenges of assuring high reliability when facing suicide terrorism, in *Seeds of Disasters*, ed. by P. Auerswald, L. Branscomb, T.R. LaPorte, E.O. Michel-Kerjan (Cambridge University Press, New York, 2006)
11. J.C. Le Coze, K. Pettersen, T. Reiman, The foundations of safety science, Saf. Sci. **67**, 1–5.
12. C. Perrow, *Normal accidents: living with high-risk technologies* (Basic Books, New York, 1984)
13. K.A. Pettersen, T. Bjornskau, Organizational contradictions between safety and security—perceived challenges and ways of integrating critical infrastructure protection in civil aviation, Safety Science vol. 71, pp. 167–177, Elsevier (2015)
14. N. Pidgeon, Observing the English weather: a personal journey from safety I to IV, in J.C. Le Coze (ed) Safety Science Research: Evolution, Challenges and New Directions, pp. 269–280, CRC Press (2019)

15. L. Piètre-Cambacédès, M. Bouissou, Cross-fertilization between safety and security engineering, Reliab. Eng. Syst. Saf. **110**:110–126, Elsevier (2013)
16. L. Piètre-Cambacédès, C. Chaudet, The SEMA referential framework: avoiding ambiguities in the terms "security" and "safety", Int. J. Critical Infrastr. Protect. **3**, 55–66, Elsevier (2010)
17. M. Power, The risk management of everything. J. Risk Finan. **5**(3), 58–65 (2004)
18. J.F. Short, *Organizations, Uncertainties, and Risk*. Westview Pr (1992)
19. I.M. Skierka, The governance of safety and security risks in connected healthcare (2018)
20. C. Smith, D.J., *Brooks, Security Science: The Theory and Practice of Security*. Butterworth-Heinemann (2012)
21. B.A. Turner, *Man-made Disasters* (Wykeham Press, London, 1978)
22. O. Waever, B. Buzan, M. Kelstrup, P. Lemaitre, *Identity, Migration and the New Security Agenda in Europe* (Palgrave Macmillan, New York, 1993)
23. W. Young, N. Leveson, An integrated approach to safety and security based on systems theory. Commun. ACM **57**(2) (2014)
24. C. Macrae, Close Calls: Managing Risk and Resilience in Airline Flight Safety. Springer (2014)

Chapter 2
The Concepts of Risk, Safety, and Security: A Fundamental Exploration and Understanding of Similarities and Differences

Peter J. Blokland and Genserik L. Reniers

Abstract When discussing the concepts of risk, safety, and security, people have an intuitive understanding of what these concepts mean and to a certain level, this understanding is universal. However, when delving into the meaning of the words and concepts in order to fully understand all their aspects, one is likely to fall into a semantic debate and ontological discussions. As such, this chapter explores the similarities and differences behind the perceptions to come to a fundamental understanding of the concepts, proposing a common semantic and ontological ground for safety and security science, introducing a definition of objectives as a central starting point in the study and management of risk, safety, and security.

Keywords Safety · Security · Risk · Foundation · Definitions · Similarities · Differences

2.1 Introduction

Risk and safety are often proposed as being antonyms, but more and more understanding grows that this is only partially true and not in line with the most modern, more encompassing views on risk and safety [1–4]. Likewise, safety and security are often seen as being completely different fields of expertise and study, separated from each other, while other views might more underline the similarities that are to be found between the two concepts and how they can be regarded as being synonyms [6].

P. J. Blokland (✉) · G. L. Reniers
Safety and Security Science Group (S3G), Delft University of Technology, Delft, The Netherlands
e-mail: P.J.Blokland@tudelft.nl

G. L. Reniers
Center for Corporate Sustainability (CEDON), KULeuven - Campus Brussels, Brussels, Belgium

Faculty of Applied Economics Sciences (ENM), Department Engineering Management,
University of Antwerp, Antwerp, Belgium

© The Author(s) 2020
C. Bieder and K. Pettersen Gould (eds.), *The Coupling of Safety and Security*, SpringerBriefs in Safety Management,
https://doi.org/10.1007/978-3-030-47229-0_2

So, how do these concepts relate to each other? How can a contemporary and inclusive view on risk, safety, and security help in understanding and in dealing with the issues related to these concepts?

2.2 The Concepts of Risk, Safety, and Security

Perceptions and awareness regarding the concepts of safety, security, and risk have evolved in recent years from a narrow and specialist perspective to a more holistic view on, and approach toward the related issues. However, this understanding is not necessarily a common perspective. The whole world comprehends what the words mean and in one's own perception how they can be understood. However, when opening a discussion on what these concepts really are, and how one should study or deal with them, it is most likely to end up in ontological and semantic debates due to the different views, perceptions, and definitions that exist.

2.2.1 The Importance of Standardization and Commonly Agreed upon Definitions of Concepts

Science, including the domain of risk and safety, largely depends upon clear and commonly agreed upon definitions of concepts, and well-defined parameters, because having precise definitions of concepts and parameters allows for standardization, enhances communication, and allows for an unambiguous sharing of knowledge. Our ability to combine information from independent experiments depends on the use of standards analogous to manufacturing standards, needed for fitting parts from different manufacturers [7].

2.2.2 Synonyms and Antonyms

When studying in the field of safety and security science, it is hard to find unambiguous definitions and parameters that clearly link safety, security, and risk. After reviewing the safety science literature, it is clear that the question "what is safety" can be answered in many ways and that it is very hard to find a clear definition of its opposite.

As a consequence, the study of the concepts of risk, safety, and security shows that there is no truly commonly accepted and widely used semantic foundation to be used in Safety and Security Science. Likewise, such a study also confirms that there is a lack of standardization when it comes to defining the opposite, the antonyms that indicate a lack of safety or security.

Table 2.1 Google Scholar search results—27 March 2018

Concept	Number of hits	Concept	Number of hits
Risk	4.770.000	Uncertainty	3.930.000
Safety	3.450.000	*Unsafety*	*8.800*
Security	3.290.000	*Unsecurity*	*40.800*
Accident	3.110.000	Insecurity	1.090.000
Incident	3.160.000	Mishap	77.500
Disaster	2.800.000	Catastrophe	899.000
Hazard	3.340.000	Danger	2.770.000
Injury	1.900.000	*Loss*	*5.810.000*

A perfect word to indicate a lack of safety would be "unsafety", although it is little used in scientific literature, as is indicated in Table 2.1.

For the antonym of security, it is even more difficult to find a commonly used word covering the subject. For example, the Oxford living dictionary defines unsecurity as "uncertainty or anxiety about oneself", "a lack of confidence". Is this what people generally think of when talking about security issues in safety and security science today? It is sensible to use the word "unsecurity" instead.

2.3   A Semantic and Ontological Perspective on Safety and Security

2.3.1   Standard Definitions

While standard definitions for safety and security are lacking, this is not so for the concept of risk. Although regarding the concept of "risk" many opinions and definitions exist, an encompassing standardized definition is available. The International Organization for Standardization (ISO) defines risk as *"the effect of uncertainty on objectives"*.

Taking this definition as a reference makes it possible to define safety and security and their antonyms in a similar, unambiguous, and encompassing way. Safety in its broadest sense could then be defined as follows: *"Safety is the condition/set of circumstances where the likelihood of negative effects on objectives is Low"* [5].

Risk and safety—where safety needs to be understood in a broad perspective including security—are tightly related and the understanding of these two concepts have evolved in similar ways, expanding the view from a pure loss perspective toward a more encompassing view, including negative (loss) and positive (gain) effects. Also in safety science, the awareness rises that the domain of safety does not only cover the protection against loss (Safety-I), but also includes the condition of excellent performance in achieving and safeguarding objectives (Safety-II) [8].

Today, risk, safety, and security are also linked to what one actually wants and how to get what one wants. However, it is this most obvious part, "the objectives", that is often forgotten in definitions, while the concept of objective is maybe the most important element in understanding the concepts of risk, safety, and security.

2.3.2   Linking and Differentiating Risk and Safety

What one "wants" can be considered as one's "objectives", with the concept "objective" understood in its most encompassing way. The following comprehensive definition of the concept "objective" is proposed as a fundamental starting point for the comprehension of the concepts risk, safety, and security: *"Objectives are those matters, tangible and intangible, that individuals, organisations or society as a whole (as a group of individuals) want, need, pursue, try to obtain or aim for. Objectives can also be conditions, situations or possessions that have already been established or acquired and that are, or have been, maintained as a purpose, wanted state or needed condition, whether consciously and deliberately expressed or unconsciously and indeliberately present".*

2.3.3   Linking Risk and Safety

As such, based on the ISO definition of risk, the link between risk and safety can be seen as follows: risk, in order to exist, requires the presence of all three of the following elements: "objectives", "effects" that can affect those objectives, and "uncertainty" related to these elements. Safety (including security) mainly concerns the objectives and the effects that can affect these objectives. Understanding risk and safety (including security) then both requires the understanding of the objectives that matter, the possible effects that can affect these objectives (*negative effects on objectives*), the likelihood of occurrence of these effects, and the level of impact of these effects and the associated likelihood (*likelihood is low*).

2.3.4   Differentiating Risk and Safety

The only fundamental difference between risk and safety consists in the fact that risk deals with an uncertain future state, while safety is more concerned with determined, actual conditions. When these effects are positive, they enhance safety as they will support the objectives involved, while the negative effects degrade safety, or increase unsafety, as they subtract from the related objectives.

2.3.5 Quality of Perception

Irrespective of the actual conditions and future possible outcomes, risk, safety, and security will always vary from one individual to another due to variations in objectives and related values. As such, risk, safety, and security are constructs in people's minds. Every individual has different sets of objectives or values the same objectives differently, creating different perceptions of the same reality.

Furthermore, every individual has their own unique perception of reality, because reality will always need an interpretation and can only be perceived. Hence, there will always be a remaining level of uncertainty and residual lack of understanding related to risk, safety, and security, varying from one person to another. Safety science should, therefore, aim for the highest possible quality of perception, where the deviation between reality as it is and the perception of this reality is the lowest possible.

2.3.6 Constraints

Pursuing or safeguarding objectives will always be accompanied by the effects of uncertainty originating from a variety of risk sources. When managing risk, aiming for safety and security, one, therefore, has to identify the risk sources and their associated risks. Pursuing and safeguarding objectives requires certain levels of risk not to be exceeded. These constraints are to be taken into account when managing risk and to be adhered to when safety is a concern.

2.4 Linking and Differentiating Safety and Security

So far, safety and security have been regarded in the same way. However, what are the common elements that make security the same as safety and what sets these two concepts apart?

2.4.1 A Distinction on the Level of "Effect"

In managing risk, risk professionals mainly try to determine the level of risk once risks have been identified. However, the assessment of the nature of risk is an important additional element to consider in managing risk and therefore, also in determining and managing safety.

The level of risk can be understood as being the level of impact of the effects on objectives (negative and positive) in combination with their related level of uncertainty. It is often expressed in the form of a combination of probabilities and consequences. The nature of risk on the other hand is more linked to the sources of risk and how these risks emerge and develop. In ISO Guide 73, a risk source is defined as being an element that, alone or in combination, can give rise to risk. It is in the understanding of possible risk sources that the difference between safety and security can be found.

When continuing on the semantic foundation provided by ISO 31000 and ISO Guide 73, safety can be seen as "*a condition or set of circumstances, where the likelihood of negative effects of uncertainty on objectives is low*". When safety is regarded in a very general way, security then is just a sub-set of safety. Indeed, when the likelihood of negative effects of uncertainty on objectives is low, this also means that a secure(d) condition or set of circumstances exists.

A first (and obvious) distinction between safety and security can be discovered when looking at the "*effects*" of uncertainty on objectives, introducing the idea that effects can be regarded as being "*intentional*" or "*unintentional*" (accidental). When negative effects on objectives are "intentional", it is appropriate and correct to use the term security instead of speaking of safety. Consequently, it would also be inappropriate to use the term "security" when the effects of uncertainty involved are "unintentional". Terrorists, as an example, intend to cause damage and harm. They intentionally increase the likelihood of negative effects on the objectives of the groups of people or parts of society they want to terrorize. Likewise, criminals intentionally act against the laws meant to safeguard specific societal, organizational, or individual objectives. Security, therefore, can be defined as follows: "*Security is the condition/set of circumstances where the likelihood of intentional negative effects on objectives is low*" [5].

2.4.2 A Distinction on the Level of "Objectives"

Another way to look at the difference between safety and security, on a more fundamental level, is to examine the concerned objectives. A typical aspect of a security setting is the involvement of multiple parties (with a minimum of two). Different perceptions come into play and accordingly also different objectives become involved. One of the parties will try to achieve, maintain, and protect a set of objectives, whereas one or more opposing parties will have different opinions on those objectives, and they may intentionally try to affect these objectives in a negative way. When looking at security situations from this perspective, it becomes clear that security issues can be regarded as situations or sets of circumstances where different, non-aligned, objectives of stakeholders conflict with each other.

If we think of objectives as vectors, pointing in a defined direction, (non)alignment of objectives can be determined in a geometrical way, and the difference between safety and security can be determined by measuring the level of non-alignment of objectives of the different parties involved. Once the non-alignment of objectives becomes more than 90° (supposing fully aligned objectives are at a 0° deviation of each other), it is apparent that these objectives are conflicting and achieving the objective of one party will cause negative effects on the objectives of the other party. Therefore, one could argue that in security management, discovering the presence of different, opposing objectives is crucial.

A level of distinction between safety and security also can be found in the level of alignment of objectives of individuals, organizations, or societies. Building on the definition of security in the preceding section and including the alignment perspective, a definition for unsecurity can be proposed: *"Unsecurity is the conditions/set of circumstances where the alignment of objectives is low and where the likelihood of intentional negative effects on objectives is high"*.

Terrorism, for example, is a very clear illustration of non-alignment of objectives, because many terrorist objectives are exactly opposite to the societal, organizational, and individual objectives they oppose.

2.4.3   A Distinction on the Level of "Uncertainty"

Last, but not least, a distinction on the level of uncertainty can be made. Safety science and safety management often depend on statistical data in order to develop theories and build safety measures. The nature of unintentional effects makes it so that the same events repeat themselves in different situations and circumstances. Furthermore, any individual can be taken into account for objectives that are very much aligned, such as keeping one's physical integrity. This provides for a vast amount of data that can be used to build theories and measures, based on statistical instruments.

Unfortunately, in security issues, the intentional nature and the non-alignment of objectives lead to repeated attempts to invent new tactics and techniques to achieve the non-aligned objectives, making it much more difficult to build on statistical data to determine specific uncertainties. It also means that different tools can and must be used to determine levels of risk and safety/security, such as game theoretical models.

The similarities and distinctions between the discussed concepts can be imagined as follows. Risk emerges when objectives are present (conscious or unconscious, deliberate or indeliberate). Risk becomes a safety or unsafety concern when objectives are tied to a specific situation or set of circumstances containing specific risk sources, providing for possible effects of uncertainty on objectives. When more than one party is involved, conflicting objectives can develop, resulting in deliberate negative effects of uncertainty on objectives for either party, making safety issues become security issues.

Safety becomes security when conflicting objectives between different parties develop, because due to the conflict, negative effects become intentional and by their intentionality also, the nature of uncertainty changes.

2.5 Conclusion

In this chapter, we have briefly described the concepts risk, safety, and security and have expounded on their similarities and differences. Subsequently, we have proposed a semantic and ontological foundation for safety and security science, introducing a definition of objectives as a central starting point in the study and management of risk, safety, and security.

References

1. T. Aven, Safety is the antonym of risk for some perspectives of risk. Saf. Sci. **47**(7), 925–930 (2009)
2. T. Aven, On how to define, understand and describe risk. Reliab. Eng. Sys. Saf. **95**(6), 623–631 (2010)
3. T. Aven, O. Renn, E. Rosa, The ontological status of the concept of risk. Saf. Sci. **49**, 1074–1079 (2011)
4. T. Aven, What is safety science? Saf. Sci. **67**, 15–20 (2014)
5. P. Blokland, G. Reniers, *Safety and Performance: Total Respect Management (TR3M): a Novel Approach to Achieve Safety and Performance Pro-actively in any organisation* (Nova Science Publishers, New York, 2017)
6. M. Boholm, N. Möller, S.O. Hansson, The concepts of risk, safety, and security: applications in everyday language. Risk Anal. **36**(2), 320–338 (2016)
7. A. Brazma, On the importance of standardisation in life sciences. Bioinformatics **17**(2), 113–114 (2001)
8. E. Hollnagel, *Safety-I and Safety–II: the Past and Future of Safety Management*. Ashgate Publishing Ltd (2014)

Chapter 3
Safety and Security Are Two Sides of the Same Coin

Nancy Leveson

Abstract Whether safety and security overlap depends on how one defines each of these qualities. Definitions are man-made and the definer can include or exclude anything they want. The question really is what the definitions imply for the solution of the problems being defined and which definitions lead to the most effective achievement of the property or properties being defined. This chapter starts by proposing an inclusive definition that combines safety and security and then discusses the practical implications of this definition for solving our safety and security problems. These implications include (1) recognizing that safety and security are not equivalent to reliability, (2) broadening the focus in security from information security and keeping intruders out, and (3) creating new integrated analysis methods based on system theory.

Keywords Safety · Security · STAMP · STPA

3.1 Definitions Are Boring But Necessary

Definitions of the terms we use are necessary for effective communications. There is no right or wrong definition, only the one we choose to use. If we limit our definition of the terms "safety" and "security", then we can effectively limit any overlap. Limited definitions, however, may also limit potential solutions to the problems. If we start from more inclusive and practical definitions, then overlap and common approaches to achieving the properties are possible.

Safety has been a part of engineering for at least 100 years and has been a concern to societies for much longer than that. Those in engineering use a precise definition of the term, while others, in social sciences, for example, tend to use much less carefully crafted definitions and sometimes change the definition depending on local context or goals. The definition of safety also differs among industries. Some limit safety and accidents to include only those events that impact on human life and injury.

N. Leveson (✉)
Department of Aeronautics and Astronautics, MIT, Cambridge, USA
e-mail: leveson@mit.edu

© The Author(s) 2020
C. Bieder and K. Pettersen Gould (eds.), *The Coupling of Safety and Security*, SpringerBriefs in Safety Management,
https://doi.org/10.1007/978-3-030-47229-0_3

Commercial aviation has historically defined safety in terms of aircraft hull losses. Some industries, such as nuclear power, which have serious political concerns, have proposed politically useful definitions but ones that are almost useless in engineering design.

The most inclusive definition, started in the U.S. Defense industry after WWII, is the one used in this chapter:

Definition Safety is freedom from accidents (losses).

Definition An accident/mishap is any undesired or unplanned event that results in a loss, as defined by the system stakeholders.

Losses may include loss of human life or injury, equipment or property damage, environmental pollution, mission loss (non-fulfillment of mission), negative business impact (e.g., damage to reputation, product launch delay, legal entanglements), etc. There is nothing in the definition that distinguishes between inadvertent and intentional causes. In fact, the definition does not limit the causes in any way. So security is included in the definition.

The concept of a *hazard* is critical in safety engineering.

Definition A hazard is a system state or set of conditions that, together with some (worst-case) environmental conditions, will lead to a loss.

Note that hazards are defined in safety engineering as states of the system, not the environment. The ultimate goal of safety engineering is to eliminate losses, but some of the conditions that lead to a loss may not be under the control of the designer or operator, i.e., they are outside the boundary of the designed and operated system. So for practical reasons, hazards are defined as system states that the designers and operators never want to occur and thus try to eliminate or, if that is not possible, control. Although the term "hazard" is sometimes loosely used to refer to things outside the system boundaries, such as inclement weather or high mountains in aviation, hazards in safety engineering are limited to system states that are within the control of the system designer. In this example, the hazard is not the inclement weather or mountain, but rather it is the aircraft being negatively impacted by inclement weather or the aircraft violating minimum separation from the mountain. We cannot eliminate the weather or the mountain, but we can control the design and operation of our system to eliminate the threat posed by the weather or mountain. Constraints or controls may involve designing the aircraft to withstand the impact of the weather or it may involve operational controls such as staying clear of the weather or the mountain. Thus, the goal of the designers and operators is to identify the system hazards (defined as under the control of the designers) and eliminate or control them in the design and operation of the system.

In security, the equivalent term for a hazard is a vulnerability, i.e., a weakness in a product that leaves it open to a loss. In the most general sense, security can be defined in terms of the system state being free from threats or vulnerabilities, i.e., potential losses. Here hazard and vulnerability are basically equivalent.

Definition <u>Hazard analysis</u> is the process of identifying the causal scenarios of hazards.

While hazard analysis usually only considers scenarios made up of inadvertent events, including security requires only adding a few extra causal scenarios in the hazard analysis process. This addition will provide all the information needed to prevent losses usually considered as security problems. For example, the cause of an operator doing the wrong thing might be that he or she is inadvertently confused about the state of the system, such as thinking that a valve is already closed and therefore, not closing it when required. That misinformation may result from a sensor failure that provides the wrong information or may result from a hostile actor purposely providing false information. These considerations result in adding during the analysis more paths to get to the hazardous state (which must be dealt with in design or operations), but do not necessarily change the way the designer or operator of the system attempts to prevent that operator unsafe behavior (see the Stuxnet example below).

Definition The goal of <u>safety engineering</u> is to eliminate or control hazard scenarios in design and operations.

The difference between physical security and cybersecurity is irrelevant except that cybersecurity focuses on only one aspect of the system design and thus has a more limited scope. Physical system security now almost always includes software components and thus cybersecurity is usually a component of physical system security.

3.2   Safety and Security Are Not Equal to Reliability

There has been much confusion between safety and reliability, which are two very different qualities.[1] When systems were relatively simple, were made up solely of electromechanical parts, and could be exhaustively analyzed or tested, design errors leading to a loss could be identified and eliminated for the most part before the system was fielded and used: the remaining causes of losses were primarily physical failures. The traditional hazard analysis techniques (which are used to identify the potential causes of the system hazards), such as fault tree analysis, HAZOP (in the chemical industry), event tree analysis (in the nuclear industry), FMECA (failure modes and criticality analysis) all stem from this era, which includes the 1970s and before. For these relatively simple electromechanical systems, reliability of the components was a convenient proxy for safety, as most accidents resulted from component failure. Therefore, the analysis techniques were designed to identify the component failures that can lead to a loss.

---

[1]It is not possible to discuss this confusion in depth in this short paper. The reader is referred to Nancy Leveson, *Engineering a Safer World*, MIT Press, 2012 [1] for an extensive discussion.

Since the introduction of computer controls and software in critical systems starting around 1980, system complexity has been increasing exponentially. The bottom line is that system design errors (i.e., system engineering errors) cannot be eliminated prior to use and are an important cause of accidents in systems today. There is also increased recognition that losses can be related to human factors design, management, operational procedures, regulatory and social factors, and changes within the system or in its environment over time. This is true for both safety and security. System components can be perfectly reliable (they can satisfy their stated requirements and thus do not fail), but accidents can (and often do) occur. Alternatively, system components and indeed the system itself can be unreliable, and the system can still be safe. Defining safety or security in terms of reliability does not work for today's engineered systems. Losses are not prevented by simply preventing system or component failures.

3.3   We Need to Broaden the Focus from Information Security and Keeping Intruders Out

Too often the focus in security, particularly cybersecurity, is on protecting information. But there are important losses that do not involve information that are for the most part being ignored. These losses involve mission assurance. The loss of power production from the electrical grid or a nuclear power plant or the loss of the scientific mission for a spacecraft is just as important (and in some respects more important) than the loss of information. In addition, given that it has proven virtually impossible to keep people out of systems, particularly cyber systems that are connected to the outside world, preventing those with malicious intentions from entering our systems does not appear to be an effective way to solve the security problem.

While intentionality does differ between safety and security, intentionality is not very important when analyzing safety and security and preventing losses. That difference is irrelevant from a safety engineering perspective when the consequences are the same. Whether the explosion of a chemical plant, for example, is the result of an intentional act or an inadvertent one, the result is the same, i.e., harmful to both the system and the environment. Intentionality simply adds some additional causal scenarios to the hazard analysis. The techniques used to identify and prevent causal scenarios for both can be identical.

As an example, consider the Stuxnet worm that targeted the Iranian nuclear program. In this case, the loss was *Damage to the reactor* (specifically, the centrifuges). The hazard/vulnerability was that *the centrifuges are damaged by spinning too fast*. The constraint that needed to be enforced was that the *centrifuges must never spin above a maximum rate*. The hazardous control action that occurred was *issuing an increase speed command when the centrifuges are already spinning at the maximum speed*. One potential causal scenario is that the operator/software controller thought that the centrifuges were spinning at less than the maximum speed. This mistake could be inadvertent (a human or software error) or (as in this case) deliberate. But no matter which it was, the most effective potential controls for both cases are the same and include such designs as using a mechanical limiter (interlock) to prevent excess spin rate or an analog RPM gauge.

Note that security concerns need not start from outside the system: Security breaches can actually start from inside the system and the results can wreak havoc on the environment.

3.4 More Effective Approaches to Safety and Security Require a Paradigm Change

Finding more effective solutions to safety and security problems requires reconsidering the foundation on which the current solutions rest, that is, the models of causality that we assume are at the root of safety and security problems. Traditionally, accidents or losses are seen as resulting from a chain of failure events, where A fails and causes the failure of B and so on until the loss occurs. This model (called the Domino or, more recently, the Swiss cheese model of accident causation) has been around a very long time, but our engineered systems are very different than those that existed previously. The model no longer accounts for all the causes of accidents today.

To find more effective solutions to safety and security problems requires a paradigm change to a model of causality based on system theory. System Theory arose around the middle of the last century to deal with the increased complexity of the systems we were creating.

A new, more inclusive model of accident causality based on system theory is STAMP (System-Theoretic Accident Model and Processes) [1]. Instead of treating accidents as simply the result of chains of failure events, STAMP treats safety and security as a dynamic control problem where the goal is to enforce constraints on the behavior of the system as a whole, including individual component behavior as well as the interactions among the system components. In the Stuxnet example, the required system constraint was to control the rotational speed of the centrifuges to reduce wear. Other example constraints might be that minimum separation is maintained between aircraft and automobiles, that chemicals or radiation is never released from a plant, that workers must not be exposed to workplace hazards, or that a bomb must never be detonated without positive action by an authorized person. STAMP basically extends the traditional causality model to include more than just failures.

STAMP is just a theoretical model. On top of that model, a variety of new (and more powerful) tools can be created. CAST (Causal Analysis based on System Theory) can be used for analyzing the cause of losses that have already occurred. The causes may involve both unintentional and intentional actions. Security-related losses have been analyzed using CAST.

A second tool, STPA (System-Theoretic Process Analysis), can be used to identify the potential causes of losses that have not yet occurred but could in the future, i.e., to perform hazard analysis by identifying loss scenarios [2]. The potential causes of future accidents identified by STPA provide information about design and operation that system designers and operators can use to eliminate or control the identified causal scenarios.

To give the reader some feeling for what is produced by STPA and how safety and security are handled in an integrated manner, consider an aircraft ground braking system. The system-level deceleration hazards might be defined as:

H-4.1   Deceleration is insufficient upon landing, rejected takeoff, or during taxiing
H-4.2   Asymmetric deceleration maneuvers aircraft toward other objects
H-4.3   Deceleration occurs after V1 point during takeoff.

The V1 point is the point where braking during takeoff is dangerous and it is safer to continue the takeoff than to abort it.

STPA is performed on a functional model of the system. An example is shown in Fig. 3.1 where the Flight Crew (humans) control the Brake System Control Unit (BSCU), which is composed of an autobrake controller and a hydraulic controller, both of which will be composed of a significant amount of software in today's aircraft. The BSCU controls the Hydraulic Controller, which actually provides the physical commands to the aircraft wheel brakes. The Flight Crew can also send commands directly to the hydraulic braking system to decelerate the aircraft.

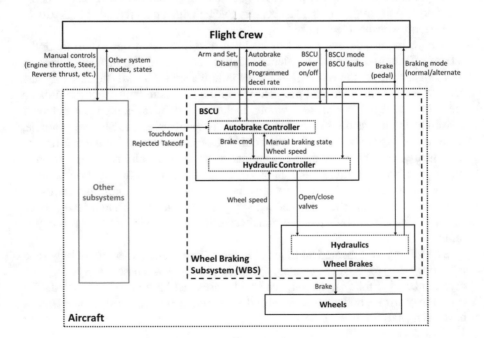

Fig. 3.1 Functional control structure of the wheel braking system

STPA is performed on this control structure. The analysis starts in the same way for both safety and security, i.e., nothing additional is needed to handle security until the end of the process. First, the potential unsafe/insecure control actions are identified. A small example is shown in Table 3.1 for the BSCU *Brake* control action. The table contains the conditions under which this control action could lead to a system hazard (H-4.1 in this partial example). Table 3.2 shows that the same process can be performed for the humans in the system, they are treated the same as any system component.

The next step is to identify the scenarios that can lead to these unsafe control actions. The scenarios will include the normal failure scenarios identified by the traditional hazard analysis techniques such as FTA, FMECA, HAZOP, but almost always more than they produce. UCA-1 in Table 3.1 is "BSCU Autobrake does not provide the Brake control action during landing roll when the BSCU is armed [H-4.1]". Because pilots may be busy during touchdown, this example braking system allows them to set an automatic braking action (autobrake) to brake after touchdown occurs. UCA-1 in Table 3.1 is that the autobrake does not activate when it has been set. The hazard analysis goal then is to identify the reasons why this unsafe control

Table 3.1 Examples of unsafe control actions for the BSCU (partial example)

Control action	Not providing causes hazard	Providing causes hazard	Too early, too late, out of order	Stopped too soon, applied too long
Brake	UCA-1: BSCU Autobrake does not provide the Brake control action during landing roll when the BSCU is armed [H-4.1]	UCA-2: BSCU Autobrake provides Brake control action during a normal takeoff [H-4.3, H-4.6] UCA-5: BSCU Autobrake provides Brake control action with an insufficient level of braking during landing roll [H-4.1] UCA-6: BSCU Autobrake provides Brake control action with directional or asymmetrical braking during landing roll [H-4.1, H-4.2]	UCA-3: BSCU Autobrake provides the Brake control action too late (>TBD seconds) after touchdown [H-4.1]	UCA-4: BSCU Autobrake stops providing the Brake control action too early (before TBD taxi speed attained) when aircraft lands [H-4.1]

Table 3.2 Example unsafe control actions for the flight crew (partial example)

Control action	Not providing causes hazard	Providing causes hazard	Too early, too late, out of order	Stopped too soon, applied too long
Power Off BSCU	UCA-1: Crew does not provide BSCU Power Off when abnormal WBS behavior occurs [H-4.1, H-4.4, H-7]	UCA-2: Crew provides BSCU Power Off when Anti-Skid functionality is needed and WBS is functioning normally [H-4.1, H-7]	Crew powers off BSCU too early before Autobrake or Anti-Skid behavior is completed when it is needed [H-4.1, H-7]	N/A

action could occur. The scenarios can be used to create safety/security requirements and to design the scenarios out of the system.

- Scenario 1: UCA-1 could occur if the BSCU incorrectly believes the aircraft has already come to a stop. One possible reason for this flawed belief is that the received feedback momentarily indicates zero speed during landing roll. The received feedback may momentarily indicate zero speed during anti-skid operation, even though the aircraft is not stopped.
- Scenario 2: The BSCU is armed and the aircraft begins the landing roll. The BSCU does not provide the brake control action because the BSCU incorrectly believes the aircraft is in the air and has not touched down. This flawed belief will occur if the touchdown indication is not received upon touchdown. The touchdown indication may not be received when needed if any of the following occur:

  - Wheels hydroplane due to a wet runway (insufficient wheel speed),
  - Wheel speed or weight on wheel feedback is delayed due to the filtering used,
  - Conflicting air/ground indications due to crosswind landing,
  - Failure of wheel speed sensors,
  - Failure of air/ground switches,
  - Etc.

As a result, insufficient deceleration may be provided upon landing [H4.1].

To include causes related to security, only one additional possibility needs to be considered: identify how the scenarios, for example, the specified feedback and other information, could be affected by an adversary. More specifically, how could feedback and other information be injected, spoofed, tampered, intercepted, or disclosed to an adversary? The following causes might be added to the scenario above to include security:

- Adversary spoofs feedback indicating insufficient wheel speed
- Wheel speed is delayed due to adversary performing a DoS (Denial of Service) attack
- Correct wheel speed feedback is intercepted and blocked by an adversary
- Adversary disables power to the wheel speed sensors.

Scenarios must also be created for the situations where a correct and safe control action is provided but it is not executed. In our example, the BSCU sends the brake command but the brakes are not applied. Some example scenarios for this case are as follows:

- Scenario 3: The BSCU sends a Brake command, but the brakes are not applied because the wheel braking system was previously commanded into an alternate braking mode (bypassing the BSCU). As a result, insufficient deceleration may be provided upon landing [H-4.1].
- Scenario 4: The BSCU sends Brake command, but the brakes are not applied due to insufficient hydraulic pressure (pump failure, hydraulic leak, etc.). As a result, insufficient deceleration may be provided upon landing [H-4.1].
- Scenario 5: The BSCU sends Brake command, the brakes are applied, but the aircraft does not decelerate due to a wet runway (wheels hydroplane). As a result, insufficient deceleration may be provided upon landing [H-4.1].

Again, to include security issues, the same additional possibilities need to be considered, i.e., identify how adversaries can interact with the control process to cause the unsafe control actions. For example,

- Scenario 6: The BSCU sends Brake command, but the brakes are not applied because an adversary injected a command that puts the wheel braking system into an alternate braking mode. As a result, insufficient deceleration may be provided upon landing [H-4.1].

STPA can handle humans in the same way it handles hardware and software. Table 3.2 shows an example of the crew responsibility to power off the BSCU. As one simple example,

Crew-UCA-1	Crew does not provide BSCU Power Off when abnormal WBS behavior occurs [H-4.1, H-4.4].
Scenario 1 for Crew-UCA-1	Abnormal WBS behavior occurs and a BSCU fault indication is provided to the crew. The crew does not power off the BSCU [Crew-UCA-1] because the operating procedures did not specify that the crew must power off the BSCU upon receiving a BSCU fault indication.

Sophisticated human factors considerations can be included here, but this topic is beyond the scope of this chapter.

3.5 What Can We Conclude from This Argument?

Safety and security can be considered using a common approach and integrated analysis process if safety and security are defined appropriately [3]. The definitions commonly used in the defense industry provide this facility. Other limitations in how we handle these properties also need to be removed to accelerate success in achieving these two properties, which are really just two sides of the same coin:

- Safety analysis needs to be extended beyond reliability analysis,
- Security has to be broadened beyond the current limited focus on information security and keeping intruders out, and
- A paradigm change is needed to go beyond accidents considered to be a chain of failure events and basing our hazard analysis techniques on reliability theory to one where our accident causality models and hazard analysis techniques are based on system theory.

Will these changes provide greater success? The system-theoretic approach to safety engineering and the related integrated approach to safety and security have been experimentally compared with current approaches many times and empirically compared by companies on their own systems. In all comparisons (now numbering a hundred or so), the system-theoretic and integrated approaches have proven superior to the traditional approaches. They are currently being used on critical systems around the world and in almost every industry, but particularly in automobiles and aviation where autonomy is advancing quickly.

References

1. N.G. Leveson, *Engineering a Safer World* (MIT Press, 2012)
2. N.G. Leveson, J.P. Thomas, *STPA Handbook* (2018) http://psas.scripts.mit.edu/home/get_file.php?name=STPA_handbook.pdf. Accessed 22 Sept 2018
3. W. Young, N.G. Leveson, An integrated approach to safety and security based on systems theory. Commun. ACM **57**(2), 31–35 (2014)

Chapter 4
Safety Versus Security in Aviation

Heinz Wipf

Abstract The two domains safety and security have traditionally been kept separated in aviation. While the first treats risks associated with aviation activities, the
latter safeguards civil aviation against acts of unlawful interference. While national
and international guidelines exist in addressing the installation of risk management
for organizations having hazardous operations in aviation, an appropriate application
of established assessment techniques, both quantitative and qualitative are crucial to
both domains. For an incorrect hazard identification and the quantification of an
adverse outcome may strongly affect both the level of protection and the investments required to reach it. The empirical example and data shown stem from safety
risk assessments in HEMS (helicopter emergency medical service) flight operations.
These flight operations use advanced instrument flight procedures in obstacle rich
environments under low visibility conditions and are therefore a safety concern on the
one hand. On the other hand, one analyzes security, whenever HEMS flights are operated in adverse weather conditions, having as a sole navigation source signals from a
global navigation satellite constellation. A traditional safety risk assessment (Wipf in
Aviation risk and safety management, Springer, p 108, 1) under these circumstances,
considers only factors of human performance under technical failure conditions. A
security analysis, however, should treat all forms of jamming, meaconing, and spoofing of the satellite signals and the adverse impact on the performance of the receiver
to calculate a valid position. The chapter illustrates to which extent commonalities
reign in both domains and where practices go separate ways.

Keywords GNSS · Air navigation · HEMS · Safety · Radio frequency
interference · Game theory

H. Wipf (✉)
Airnav Consulting Zurich, Zurich, Switzerland
e-mail: airnavconsulting@bluewin.ch

© The Author(s) 2020
C. Bieder and K. Pettersen Gould (eds.), *The Coupling of Safety
and Security*, SpringerBriefs in Safety Management,
https://doi.org/10.1007/978-3-030-47229-0_4

4.1   Introduction

Over the last years based on our experience with light helicopter operations for disaster relief, search and rescue, and Helicopter Emergency Medical Services (HEMS), the necessity of an ever-widening operational scenario with all-weather capability has become apparent.

The use of Global Navigation Satellite Systems (GNSS) as a primary navigation source under low visibility conditions was, therefore, obvious. Due to weight restrictions and topographical circumstances, these signals often are the only means of getting a position solution. The relevant signals containing navigation information allowing the receiver to estimate the position are transmitted over an openly accessible radio frequency channel. Propagation effects [2] induced by flight attitude in conjunction with the receiver's antenna pattern may impair the quality of the navigation solution. Moreover, such a channel is prone to noise and interference stemming from different radio sources (Fig. 4.2). If such transmissions are intentional, then one can classify it as an unlawful interference. So while the former are safety-related, the latter is a security issue (Fig. 4.3).

The two domains safety and security have traditionally been kept separated because the International Civil Aviation Organization (ICAO) published different definitions in their annexes to the Chicago Convention. In these documents, security is defined as "Safeguarding civil aviation against acts of unlawful interference", while safety is "The state in which risks associated with aviation activities, related to, or in direct support of the operation of aircraft, are reduced and controlled to an acceptable level". While security is handled by entities like law enforcement agencies and airports, safety is said to depend on personnel, procedures, and equipment, which is foremost the field of air operators and air navigation service providers. Another view on this separation comes from applying Systems Engineering (SE) methods. An approach is shown in Fig. 4.1.

The SE philosophy is quite in line with the saying that hazards lead to safety incidents in the same way that vulnerabilities lead to security incidents. The same view in a more formalized arrangement is shown in Table 4.1.

Fig. 4.1 Context of safety and security from a systems engineering viewpoint

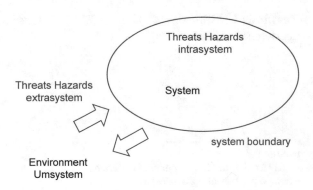

Fig. 4.2 Radio Frequency (RF) channel with noise and non-intentional interference

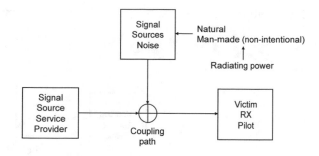

Table 4.1 Threat matrix formalizing the context in Fig. 4.2

Attacker	Victim	
	System	Environment
System	NA	Safety
Environment	Security	NA

This formalization of the 2 by 2 threat matrix in Table 4.1 reveals two entries (NA) that remain unaddressed. When asked what the synergies are between the two, the author would rather rephrase the question as: What are the commonalities? The question brought up here will be whether the two fields have to be treated differently or whether their unification is thinkable, notwithstanding the fact and existence of different authorities and jurisdictions.

4.2 The Economic Good in Question

The good in question is a radio frequency channel. The practical example chosen is at the same time relevant and valid, due to the fact that satellite navigation signals are extensively used for all sorts of critical infrastructure and hazardous operations.

Such a channel can be characterized by simple metrics, the bandwidth and the signal-to-noise ratio [3]. For this example, we would extend this ratio to also include any interfering signal power. The metric would then be signal-to-(noise + interference). The block schematic in Fig 4.2 is for the situation where only machines or one operator are present $N = \{0, 1\}$.

The system bandwidth is largely given by the base-band signal, and any interference being natural or man-made is, to the first order, only relevant within this channel bandwidth, because the receiver (RX) will band-pass all signals and suppress the others. This also means the interference format has to match the bandwidth to be effective.[1] In the case of meaconing and spoofing, this is per se the case, because the signal used to interfere is identical to the original one (Fig. 4.3).

---

[1]Here, we limit ourselves to one possible interference scenario: in-channel interference.

Fig. 4.3 RF channel with
intentional interference

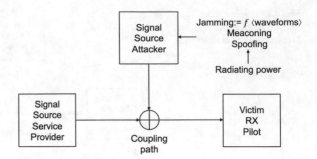

So the only free variables for the interferer are the duration and the radiated power. Although a jamming attack has the freedom of different signal formats, the classes are limited to only four.[2]

Jamming is the emission of radio frequency signals of sufficient power and with such characteristics to prevent the receivers from working properly.

Meaconing is the reception, delay, and rebroadcast of a signal with a larger power than received. At the receiving antenna, the wanted and unwanted signals are added to confuse the system. Ground- and space-based augmentation radio links could also be prone to meaconing, especially if the correct differential signal is suppressed with a stronger one containing false corrections.

Spoofing is a technique to cause a receiver to lock onto legitimate-appearing false signals. The attack will inject misleading information and thereby eventually even control the flight [4, 5, p. 63].

The radiated transmission power is a continuous variable that the attacker is free to choose for each attack. However, as indicated above, certain bounds exist. Every RF channel is specified in five dimensions. So out of frequency, time, space, modulation, only polarization would remain an issue for an optimization on the side of the interferer.

An air navigation service provider, supporting hazardous flight operations, has to inform the user of three probabilities

1. Reliability[3]: using the service and not losing it.
2. Availability: requesting the service and getting it.
3. Integrity: correctness of the information supplied.

The above include the condition that the provided signals are within specified error bounds in space and time.[4]

The constant presence of interference from natural sources is an important aspect. So even in the absence of man-made interference, the receiver has to cope with noise from intra- or extra-system sources.

---

[2]Continuous Wave (CW), chirp, pulses, and noise.

[3]Also Continuity of Service (CoS).

[4]See ITU definitions.

Another aspect indicated by the signal-to-(noise + interference) ratio is the diminishing of the signal power due to an increase in radio path attenuation. These two factors are relevant when discussing game-theoretic approaches, namely in the absence of an attacker $N = \{0, 1\}$.

4.3   A Game-Theoretic Approach Put to Practice

The title of this section reads like a contradiction in terms, but it is well worth to attempt to get practical. Game theory, a branch of mathematics, offers an analytical approach to situations of a practical nature. The situations considered are games with different parties having common or different interests. Mathematical solutions are possible for certain cases. The situations include true games as such[5] as well as real-world problems in politics, economics, or warfare. The theory has also recently been applied to terrorism [4, p. 198].

This contribution treats a real-world problem, and classical game theory is being tried. It means players may strategize,[6] decide, and act. Whereby chance, hidden or incomplete information are pertinent circumstances. A game consists of players[7] (individuals/organizations), strategies (a plan, objectives, decisions, and actions), situations, and a gain from participating (utility). In short, a theory of mathematical models is applied to formalize interdependent players with their decisions and actions under a condition of conflict or cooperation.

Thus, the question is what are the provisions of such an approach to safety and security and what are the elements necessary to model the chosen real-life situation. Elements in this example are discrete and can, therefore, be described in a set-theoretic way. The only exception is the radiated power P of the interferer. If an attacker intends to maximize impact while minimizing the probability to be detected, then this value is bounded. This parameter, therefore, is also accessible to set theory. So let the radiated power be $P = \{0, P_{\max}\}$. The two values are then equivalent to abstain or execute an attack.

4.3.1   The Players

The complete setup includes three players with different coalition aspects summarized in Table 4.2.

Although a coalition of interest exists between the user and the service provider, it may not be strong enough to have the service provider actively taking part in the

---

[5]E.g., card games or chess.

[6]To have a plan of what to accomplish, while taking intentions of other involved parties into account.

[7]For completeness, it is advisable to attribute participation and interest of the players.

Table 4.2 Players attributes

		Service-related	Coalition	
			Participation	Interest
Players	User (U)	True	True	
	Provider (P)	False	True	
	Attacker (A)	True	True	

game. The reason lies in important investments like upgrading or replacing space-based assets. Such actions would have a negative impact on service provider's utility, which is cost versus the number of users. Thus, the service provider is excluded.

4.3.2 Available Strategies

The course of action or possible strategies in this example form finite sets (SA and SU). The setup of the game has one attacker (A) and one victim, the user of GNSS (U) in a flight under low visibility condition (IMC), under Instrument Flight Rules (IFR) with no redundancy in navigation. The attacker intends to deny the use of this only system. This situation asks for an offensive strategy on the side of the attacker and a defensive one on the side of the user. The attacker has three distinctive but feasible attacks or strategies, and they constitute a finite set:

$$SA = \{Jamming, Meaconing, Spoofing\}.$$

For the location of the jammer, different options exist. It could be on a fixed, ground-mobile, or airborne platform. We limit our case to the fixed option. Although a mobile jammer would be more difficult to detect, target jamming an airborne asset would be more of a challenge, since the road network would not be coincident with the projection of the victim's flight path. An airborne jammer finally would offer a number of attacking advantages, but operating costs would be considerable, to be effective. Moreover, detecting and locating the attacker would be fairly simple. The set of strategies of the attacked U[8] on the contrary is a purely defensive set:

SU = {spectrum/signal monitoring, reducing the coupling between receiving antenna and attacker's transmission, minimizing the exposure time}.

---

[8]P is only indirectly affected by the attack unless his assets are impacted. U has little influence on P to, e.g., motivate an increase in transmitting power, which would increase his signal/noise.

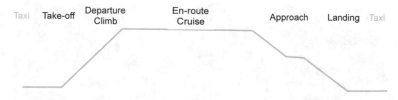

Fig. 4.4 Phases of flight after [6]

Table 4.3 FP, T_{exp}, height to the victim's antenna contrasted with criticality

FP	T_{exp}	Unit	T_{exp} h	Height in m	Criticality
Takeoff	10	sec	2.78E-03	5.0E + 01	High
Departure	5	min	8.33E-02	2.0E + 02	Medium
En-route	45	min	7.50E-01	2.0E + 03	Low
Approach	5	min	8.33E-02	2.0E + 02	Medium
Landing	30	sec	8.33E-03	2.0E + 01	High

[a]In principle from [6]

4.3.3   The Situations

The situations are governed by the phase of flight and its need for a precise aircraft position. The user counts on the three probabilities (1., 2., 3.) above indicated by the service provider. These are estimated from empirical failure rates[9] or reliability calculations. Together with corresponding exposure times, it results in failure probabilities. Figure 4.4 defines the general Flight Phases (FP).

A FP is ended and another started as decided from the flight deck (decision instance, player A). Possible scenarios, therefore, are determined and finite. Although a loss of a position solution in low visibility on ground is not irrelevant, ground movements are discarded for the sake of simplicity. The set is consequently reduced to FP = {Takeoff, Departure, En-route, Approach, Landing}.

Exposure times vary considerably. Table 4.3 shows typical mean values for helicopter operations. While T_{exp} shows changes of the order of a magnitude along the flight trajectory, the distance and with it the radio path attenuation for a potential interfering source toward the victim's receiving antenna also change.[10]

There is an intrinsic relation between exposure time and the height of the victim above the antenna of a potential interferer. This relation allows some ground for an operationalization of the probability of losing a position due to an interferer located on the ground while executing a specific flight phase. The risks for the victim depend on the status of the signal received. If the signal is in use and a critical flight

---

[9]The rates for rare events are assumed to be Poisson distributed and have an exponentially distributed duration.

[10]Path attenuation $a = 1/r^2$.

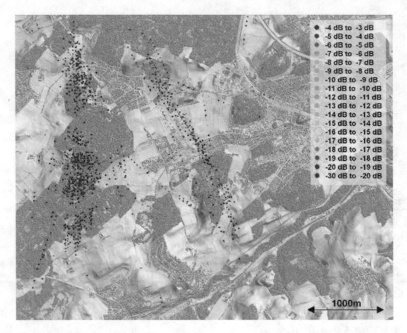

Fig. 4.5 Result of a monitoring action [7] (Figure courtesy of M. Scaramuzza, Skyguide. Included with the permission of the author.)

phase being flown, then the loss of the signal leads to a hazardous situation and the risk of an accident. If the signal is to be acquired but not available, then the mission will be aborted and economic loss results. The attacker may of course choose the interfering power[11] at his discretion. However, radiating too much power increases the Probability of Intercept (POI). This condition in turn increases the possibility of being detected by some monitoring processes [7–9]. Figure 4.5 shows the result of monitoring recorded during normal helicopter missions where the Quality of Service (QoS) is repeatedly degraded. The colors indicate the severity of potential Radio Frequency Interference (RFI).

If detected, the victim will initiate an evasive action rendering futile the attempted attack. Moreover, detection could lead to getting located by an authority in charge, so the attacker has to make a tradeoff.[12] However, the maximum radiated power of the interferer is not only bounded for tactical reasons but also for technological ones. Table 4.4 underlines the risks of an attack.

---

[11] Effective Isotropic Radiated Power (EIRP).

[12] There are more elaborate strategies in the fundus of the electronic warfare arsenal.

Table 4.4 Attacker's costs on equipment and the probability of intercept

Attack	Investment €	Knowledge	POI
Jamming	1000	Low	High
Meaconing	10,000	Medium	Low
Spoofing	1,00,000	High	Low

4.3.4   The Outcomes of the Game

The outcomes must illustrate potential gains in the areas of cost, risk, and utility. It is possible, though to include the cost in the risk for both players. The risk R for the attacker may be approached in the following way $R = (I + K) \bullet$ POI, where I is the investment for the equipment, K is the knowledge, and $I + K$ the total cost. POI is the detection of a monitoring instance within the interfered region. An attempt for the payoff matrices of the two players (A and U) is shown in Table 4.5, where the gain (1) and loss (-1) are indicated in each entry.

In this example obviously, the gain of the attacker A is the loss of the attacked U. The gain matrix above suggests a strategic advantage to attack. However, the matrix does not display the entire picture. Table 4.6 gives an indication of the likelihood that the attack is being detected and consequently a flight operational action is initiated.

The likelihood of being detected is about two orders of magnitude smaller for meaconing and spoofing compared to jamming due to the difference in signal formats.

Table 4.5   Gain matrix

Attacker \ Attacked	No action		Climb		Accelerate		climb and accelerate	
Jamming		-1	1		1		1	
	1			-1		-1		-1
Meaconing		-1	1		1		1	
	1			-1		-1		-1
Spoofing		-1	1		1		1	
	1			-1		-1		-1

Table 4.6   Likelihood of the attacked gaining situational awareness due to detecting an attack

Attacked	Attacked
	Likelihood of situational awareness
Jamming	High
Meaconing	Low
Spoofing	Low

Table 4.7 Game classification

Players	Action domain	Game type	Approach	Example
0	Safety	Non-strategic	Descriptive mathematics	Automata
1			Optimization	Socio-technical systems
2	Security	Strategic	Game-theoretic	Competition
≥3				Cooperation

In general, technical infrastructures providing a common good, accessible to the general public, are seldom attacked. An explanation may be that the attacker or his allies need the service they intend to impair for their own purposes.[13] There is a generally accepted utility attached to this good.[14] In this case, the payoff matrix must be modified to reflect such situations and to find the Nash-equilibrium [10, p. 286], which could give an explanation for this phenomenon.

4.3.5 Game-Theoretic Classification

To summarize and make use of game theory, an attempt is made to classify the example at hand. Games are classified according to the different sets mentioned above. The most obvious one is the number of players. A game can have one, two, or n players. Each manifestation has its own distinctive features, and the players need not be individuals. It may be a group of persons with common interests being part of some organization. Even organizations could federate in a game. The possibilities are summarized in Table 4.7.

The empty and the unit set of players are included to propose a possible unified approach under game-theoretic aspects. The empty set (no players) would be a purely machine-to-machine interaction, unless artificial intelligence is actively involved. The unit set (1 player) is also called a one-person game. With no rivals, the player only needs to list available strategies so to choose an optimum outcome.

When probabilities are involved, it may turn out to be more complicated. Ways and means to cope with such problems are laid down in decision theory. Or as often said, the single player is engaged in a game against nature, where nature is indifferent to the player's decision.

Whether the objectives of the players coincide or conflict is another aspect of the classification. Constant-sum games show an entirely conflicting situation (pure

---

[13] See also today's conflict zones, where mobile telephone base stations keep on working, although used for warlike actions.

[14] See physical attacks on single aircraft, but not on the infrastructure supporting flight, like vulnerable assets of air navigation or airport services.

Table 4.8 Game-theoretic classification for the example

	This example	Bi-matrix game	Matrix game
No. of players N	2	2	2
Non-cooperative	True	True	True
Finite	True	True	True
Zero-Sum	True	False	True
Strategic	True	True	True

competition),[15] with no communication between the adversaries. This fact leads to incomplete information on both sides.

Whether a game is called finite depends on finite sets above [10, p 286]. Moreover, the game cannot have an indefinite duration. In practice, there exists a window to act.

A finite non-cooperative game between two players is called a bi-matrix game. It is specified by two matrices $A = \|a_{ij}\|$ and $U = \|u_{ij}\|$ of the same dimension $m \times n$. These two matrices represent the payoff matrices (gain matrices) of the players. The strategy of player A is the selection of a row, that of player U the selection of a column. Let player A choose i ($1 < i < m$), while player U chooses j ($1 < j < n$), their respective payoffs or gains will be a_{ij} and u_{ij}. If $a_{ij} + u_{ij} = 0$ for all i, j, then the bi-matrix becomes a matrix game. The two candidates reflecting this example are either a bi-matrix or a matrix game. Table 4.8 indicates that the latter matches the situation.

4.4 Conclusion

An attempt has been made to structure a real-world problem to make it accessible for a game-theoretic solution and it appears as if the two aspects of safety and security can be assessed in one single unified solution space. Both fields turn out to be different subsets of a more fundamental superset. More formally, one rationalizes the synergy between safety and security solely in the number N of involved instances or players. So there is a temptation to see game theory as a possible means of offering a unifying approach.

A pertinent question has come into focus, namely why vulnerable basic infrastructure like radio channels in the case of air transportation has been so seldom the target of elaborate electronic attacks. One possible answer is the utility it has for all conflicting parties. In the case of openly accessible radio channels, the utility may even extend to gather information about the adversary.

---

[15]Rolling dice is an example, because the combined wealth of the players remains constant, although the distribution in the course of the game changes.

4.5   Outlook

There are other situations in aviation where game theory seems an appropriate way
to model other interactions, namely flight operators, air traffic service providers, and
airports. A typical example where airport security is negatively influencing professionals concerned with flight safety is described and analyzed in [11]. Unlike the
non-cooperative nature of the example above, these entities are engaged in a coalition game, because they have the opportunity to collaborate for mutual benefit in
several ways. Moreover, it would be advantageous to industry if rule-making and
supervisory activities would be included in such models.

References

1. H. Wipf, Risk management in air traffic control—operators risk back to basics. Aviation risk
   and safety management, Springer (2014)
2. A. Geiger et al., Simplified GNSS positioning performance analysis. Monterey ION-IEEE-
   PLANS (2014)
3. C.E. Shannon, A mathematical theory of communication. Bell Syst. Techn. J. **27**, 379–623
   (1948)
4. B. Golany et al., Nature plays with dice–terrorists do not: allocating resources to counter
   strategic versus probabilistic risks. Eur. J. Oper. Res. **192**(1)
5. Vulnerability assessment of the transportation infrastructure relying on the global positioning
   system. In: 2001 Final Report J. A. Volpe National Transportation Systems Center
6. Statistical summary of commercial jet airplane accidents worldwide operations. Seattle Boeing
   (2016)
7. M. Scaramuzza, Localization of GNSS RFI transmitters using digital surface models. Belgrade
   IFIS (2016)
8. M. Scaramuzza et al., GNSS RFI detection—finding the needle in the haystack. Tampa GNSS-
   ION (2015)
9. M. Scaramuzza et al., RFI detection in Switzerland based on helicopter recording random
   flights. Oklahoma IFIS2014 (2014)
10. J.F. Nash Jr., Non-cooperative games. Ann. Math. **54**, 286–295 (1951)
11. K.A. Pettersen et al., Organizational contradictions between safety and security perceived
    challenges and ways of integrating critical infrastructure protection in civil aviation. Saf. Sci.
    71,167–177 (2015)

Chapter 5
Security and Safety Culture—Dual or Distinct Phenomena?

Sissel H. Jore

Abstract The commission that investigated the terrorist attacks against the Algerian oil facility In Amenas concluded that the Norwegian petroleum company Statoil should establish a security culture distinct from its safety culture. Both are elements of organizational culture, so how should organizations relate to this new concept of security culture? This chapter explores the adequacy of the concept of security culture and explores whether these phenomena should be considered as a duality or separately. The adequacy of security culture is discussed in terms of how the concept is used in the In Amenas investigation report. Despite the lack of demarcation and operationalization of the security culture concept, we conclude, there is a need to further develop security culture as a theoretical and practical element. Security and safety culture should be understood separately, but in practical reality should not be treated as distinct.

Keywords Security culture · Security · Safety · Concept adequacy · Terrorism

5.1 Introduction

On 16 January 2013, the largest terrorist attack in the history of the oil and gas industry occurred at the Algerian oil facility in In Amenas; 32 heavily armed terrorists attacked the operation site, where almost 800 workers were present. Many were taken hostage in a siege that lasted four days in the middle of the Algerian desert. Terrorists killed 40 people from 10 countries, including five Statoil employees.

In the aftermath, Statoil appointed a commission to determine the relevant chain of events in the attack and to enable Statoil to improve their security, risk assessment, and emergency preparedness. The investigation report concluded that Statoil had established a security risk management system, but the company's overall capability and culture needed strengthening in order to respond to security risks associated with volatile and complex environments. The report described security culture as

S. H. Jore (✉)
University of Stavanger, Stavanger, Norway
e-mail: sissel.h.jore@uis.no

C. Bieder and K. Pettersen Gould (eds.), *The Coupling of Safety and Security*, SpringerBriefs in Safety Management,
https://doi.org/10.1007/978-3-030-47229-0_5

one important explanatory factor behind the attack's outcome and an important tool for improving security [10].

For many companies, malicious threats such as terrorism constitute a new setting. The management of such threats is often referred to as "security", in contrast to "safety", which refers to the management of risks not committed by actors with an intention to harm [6]. Along with the new responsibility for security in the private sector, new management concepts and tools aimed at guiding organizations in fulfilling this role have emerged, such as security risk management systems, security risk analysis, and security culture. The common denominator of these concepts is that they all have their counterpart in safety management, and are now being adopted and applied to the security domain. However, transferring concepts to a new area is not necessarily unproblematic. Compared to safety, security is a relatively new academic field, and "security culture" is a term seldom found in the literature [5]. Nonetheless, the recommendation of Statoil's investigation report after the In Amenas attack has led to a heightened focus on security culture in the petroleum sector. According to a 2015 study, half of the Norwegian petroleum companies included in the sample actively applied security culture as a means of security improvement [7].

Both safety and security are elements of organizational culture, so how should organizations relate to this new concept of security culture? This chapter explores the adequacy of the concept of security culture. What relationship exists between safety culture and security culture, and should these phenomena be considered as a duality or separate? The adequacy of security culture is discussed in terms of how the concept is used in the In Amenas investigation report. The criteria of conceptual goodness proposed by Gerring [2] are applied to the concept adequacy of security culture.

5.2 Distinctions Between Safety and Security

If security culture should be seen as distinct from safety culture, there is a need to investigate the content of the safety and security domains as well as the interfaces between them. In everyday use, the words "safety" and "security" invoke associations of freedom from threats and harm. Despite often being treated as synonymous, the two concepts also have diverse meanings. Frequently, the concepts are utilized to distinguish between the management of hazards from non-malicious intent (safety) and the management of threats stemming from rational humans with a malicious intent, such as sabotage, hacking, or terrorism (security) [6].

It is malicious intent that distinguishes safety from security, and not intentionality because intentionality also plays a part in safety. The literature on organizational safety has long acknowledged that accidents are neither arbitrary nor random but rather a result of insufficient resources, organization, and planning. According to these perspectives, human intent sometimes plays a role in causing accidents, and organizations should design robust measures that take into account the fact that workers sometimes intentionally diverge from standard procedures. This implies that criminal activity is not just related to security. Safety also includes rational actors that deliberately disobey rules by abusing drugs, for example, or not using safety equipment. Thus, neither intentionality nor crime is sufficient to distinguish safety from security. The difference then should be the malicious intent of the actor who actually plans to cause harm.

In contrast to safety, security is often jeopardized by external threats that most often are beyond the capability of organizations to fully know and handle. Moreover, such risks are not linked as directly to the economic profit and production system as safety risks. This means that, even though a company may have an optimal security culture, it can still be the target of a terrorist attack and experience major damage. Since a hostage crisis or terrorist attack is an extremely low-probability security event—is it really meaningful to apply the concept of culture to such extreme events in the same way as for safety?

5.3 The Investigation Report's Account of Security Culture

Statoil's investigation report concluded that Statoil had not developed a culture in which it was generally recognized that security was the shared responsibility of all and that a holistic approach to the management of security was lacking. Security was neither established as a corporate function independent of safety nor recognized for its distinctive characteristics. Furthermore, the commission claimed that, along with a lack of management commitment, security was generally not well understood throughout the organization. Thus, the ability to understand and respond to changes in the environment was characteristic of companies with a strong security culture, which would share the following characteristics (Table 5.1).

The way the commission uses the concept of security culture and their recommendations correspond with current understandings of how a security culture should be achieved. However, the recommendation to build a security culture distinct from safety might be more problematic than it seems. If organizations should put resources

Table 5.1 Characteristics of a strong security culture [10]	• Hands-on security leadership with access to top management and an ability to drive the security agenda throughout the business
	• High and clearly articulated ambitions for their security capability, which is treated as a discipline distinct from safety with clear objectives and dedicated professionals
	• Sufficient capacity and competence to identify and match the security challenges faced by the business
	• A holistic approach to the management of security risks as an integrated part of core business processes and deliveries and
	• Transparent, inclusive, active, and authoritative risk management processes run by an organization capable of identifying and acting on potential threats

Table 5.2 Criteria of conceptual adequacy based on Gerring [2]

Familiarity	How familiar is the concept to different audiences?
Resonance	Does the chosen term resonate?
Parsimony	How concise are the term and its list of defining attributes?
Coherence	How internally consistent are the instances and attributes?
Differentiation	How differentiated are the instances and attributes from similar concepts?
Depth	How many accompanying properties are shared by the instances under definition?
Theoretical utility	How useful is the concept within a wider field of inferences?
Field utility	How useful is the concept within a field of related instances and attributes?

into building a strong security culture, such programs should be based on a scientific foundation. So how does the concept of security culture stand up to scientific scrutiny?

5.4   Conceptual Adequacy of Security Culture

According to Gerring [2], concepts are critical to the functioning and evolution of science. He argues that conceptual adequacy should be perceived as an attempt to respond to the criteria delineated in Table 5.2.

5.5   Familiarity and Resonance

Security today incorporates more than just technical solutions and physical protection; it entails the management of threats from rational, strategic actors. Thus, a component that involves perception, shared understanding, and management of threats seems like a favorable contribution to the field. For this reason, the concept of security culture appears to be a promising tool for enhancing corporate security.

The degree to which a new concept "makes sense, or is intuitively" clear depends critically upon the degree to which it conforms, or clashes, with established usage in everyday language and within a specialized language community [2]. Safety culture is a well-established concept familiar to laypeople, professionals, and academics. Statoil's investigation report described security culture as a common set of beliefs, attitudes, practices, and behaviors perceived, internalized, and shared across geographic units and levels [10]. This definition corresponds to how organizational culture, safety culture, and security culture are often defined [5].

The term "safety culture" was first used as an explanatory factor in the investigation following the Chernobyl accident in 1986 [1]. Since then, the phenomenon of safety culture has been regarded as being crucial for preventing accidents in multiple sectors. Although the term and methods of measuring and achieving it remain contested, the concept is widely accepted and applied as a contributing factor to the safety of organizations [4]. By drawing on the connotations of this well-established concept, the concept of security culture suggests that security can also be achieved by applying the same tools. Thus, the concept of security culture is familiar and resonates well.

5.6 Parsimony, Coherence, Differentiation, and Depth

Even though the investigation report's definition, understanding, and recommendations on security culture align with existing theoretical perspectives in the field, the concept is new to the academic literature. The literature on security culture is minor compared to the enormous amount of research and diverse perspectives on safety culture, and is mainly developed within the nuclear, chemical, and aviation industries, building on existing theories in safety culture research. With a few exceptions, little has been written on how to achieve optimal security culture. Consequently, security culture lacks clear indicators or attributes, is poorly defined and operationalized, and lacks research linking security to organizational performance [8].

Existing definitions of security culture are tied to a specific sector, and often to a specific threat. Most of the literature that uses the term deals with information security, and not sabotage or terrorism. Thus, the literature does not take into account the polysemy of the security field. This means that security culture is not an overarching phenomenon that covers all possible security threats. Although all security threats are a crime, their *modus operandi*, target selection, and motivation differ widely and need to be addressed in very different ways.

Perhaps the greatest challenge with the term "security culture" is how it relates to and differentiates from similar terms such as "organizational culture" and "safety culture". The majority of the academic literature that uses the term argues for a holistic perspective and claims that it should be seen as an integrated part of safety culture. When attributes of security culture are defined, they are often described in the same manner as safety culture, and the specific characteristics of security are not considered. This makes it hard to differentiate the concepts from each other, both theoretically and practically. Theoretical perspectives that deal with security culture do not account for the relationships between organizational, safety, and security cultures and do not answer the fundamental questions of whether security culture is a subculture of a safety or organizational culture and what relationship exists between them.

For companies in the petroleum industry, the distinctions and overlaps between safety and security culture are of real consequence. According to the Norwegian Petroleum Framework Regulation, companies are obliged to build a safety culture ([11], Sect. 15). If security culture is seen as a subculture of safety culture, this implies that companies are also legally obligated to build a security culture.

5.7 Theoretical and Field Utility

Concepts are the building blocks of all theoretical structures, and the formation of many concepts is legitimately theory-driven. How is the concept of security situated within the broader science of security? The security field, with a few exceptions, lacks theories on organizational security, and most of its literature covers normative theories on how to achieve security without building on research because security has traditionally been connected to the Military and the Police and this been considered classified material. The concept of security culture is seldom used, and the academic literature that outlines the core elements of security science does not mention security culture [9]. Consequently, no studies describe how to establish a security culture and how security culture should be situated within the broader context of cooperate security. Thus, security culture could be a promising contribution to the literature since security science has been turning toward softer measures such as awareness, mindfulness, and resilience. A concept such as security culture could thus function as a uniting concept for how to conduct corporate security management. However, for the concept to be useful in theory formation, there is a need for both theoretical developments based on empirical studies about the role of security culture and how security culture affects security in organizations.

To enhance theoretical development, it is important to establish relationships with neighboring terms, such as safety culture. When the definitions heavily overlap, phenomena become hard to distinguish, and the newer literature attempting to operationalize the concept uses the same descriptions, attributes, and indicators [12]. Thus, there is a need to articulate the overlaps and boundaries of the concepts. Additionally, it is necessary to investigate whether a good safety culture is a prerequisite for a good security culture, and vice versa.

In practical reality, every organization has a culture (or series of subcultures) that can be expected to impact safety and security. However, it is not necessarily beneficial to simply transfer theories and concepts from the field of safety to that of security. Many aspects of the term's usage are not directly transferable to a security context. What does, for example, a "just culture" mean in the context of security when the attacker has malicious intent? This affects the possibility of transparency and openness outside trusted communities.

The investigation report states that Statoil should have a culture where every employee is dedicated to security, but is this possible or even desirable? Does a security culture mean being suspicious of colleagues and others, and is mistrust not counter to the creation of a safety culture? Detection and learning from weak signals is also problematic when perpetrators are strategic and uninterested in revealing their plans.

Theoretical discussions are often arcane and best kept within scholarly communities; however, the adequacy of the security culture concept has relevance for practical corporate security because security culture is currently not only a theoretical term but also a pragmatic tool implemented in multiple petroleum companies after the publication of the In Amenas investigation report. A study examining Norwegian petroleum companies' usage of the term concluded that although half of the companies in the study used the term, that usage did not seem to directly influence how these organizations organized their security system. The companies that rejected the security culture concept claimed this rejection was due to the difficulty of separating security from safety culture [7]. Thus, there is a practical need for security culture to be operationalized. However, we already know from safety research that a culture in an organizational context covers almost everything an organization does; thus, it becomes hard to measure the culture's impact on safety [3]. This also applies to security culture. In safety research, there is also a discussion of the relationship between culture and what is actually done in the organizations. The fact that the concept of safety culture is contested makes the transfer of perspectives from the safety to the security realm difficult.

5.8 Conclusions

There is undoubtedly a need for the concept of security culture in today's threat landscape. In complex and volatile environments, such as In Amenas, companies should implement systems that generate awareness of external threats and provide ways to handle them. For such threats, clear tactical warnings with specific information on where, when, and how a potential adversary may attack will seldom occur. This means that organizations should strive to build resilience against multiple threats, including low-probability security scenarios. These are all arguments for building a strong organizational culture with a collective security mindfulness that seeks out weak signals and strives for resilience.

Although the concept of security culture might superficially seem like a promising trajectory, the operationalization and demarcation of the concept of safety culture are so imprecise that the use of the concept may be counterproductive. However, the same can be said about the concept of safety culture, so this is not an argument for rejecting the concept.

From a practical point of view, an organization needs to deal with both safety and security risks; both influence the organizational culture. Thus, from a practical perspective, there is a need to see these concepts as a duality and not as separate phenomena. Security threats have different dynamics than safety risks, and thus security is often neglected in organizations. This is the advantage of the concept of security culture: it makes security a priority and shared responsibility.

As digitalization across industries increases and more digital assets connect to the Internet, organizations will have to increase their focus on security threats where the cultural component will play an important role, since technical solutions will be insufficient. Thus, given the increased focus on security threats in organizations today, there is a need to further develop security culture as a theoretical and practical element. Consequently, security culture could be a promising addition to the existing literature because security science is turning toward softer measures such as awareness, mindfulness, and resilience—all of which are important components of security culture.

References

1. S. Antonsen, *Safety Culture: Theory, Method and Improvement* (CRC Press, 2017)
2. J. Gerring, What makes a concept good? A criterial framework for understanding concept formation in the social sciences. Polity **31**(3), 357–393 (1999)
3. C. Gilbert, B. Journé, H. Laroche, C. Bieder (eds.), *Safety Cultures, Safety Models: Taking Stock and Moving Forward* (2018)
4. A. Hopkins, Studying organisational cultures and their effects on safety. Saf. Sci. **44**(10), 875–889 (2006)
5. S.H. Jore, Security culture—a sufficient explanation for a terrorist attack? in *Risk, Reliability and Safety: Innovating Theory and Practice: Proceedings of ESREL 2016* (CRC Press, 2017a), pp. 467–474. ISBN 9781138029972
6. S.H. Jore, The conceptual and scientific demarcation of security in contrast to safety. Eur J Secur Res, 1–18 (2017b)
7. C.I. Larsen, C. Østensjø, Operatørselskapene i petroleumssektoren sitt syn på sikringskultur: Bruk eller ikke bruk av begrepet sikringskultur. Master thesis, University of Stavanger, Norway, 2015
8. J. Malcolmson, What is security culture? Does it differ in content from general organisational culture? Paper presented at the Security Technology, 2009. 43rd Annual 2009 International Carnahan Conference on Security Technology, IEEE, 361–366 (2009)
9. C. Smith, D.J. Brooks, *Security Science: The Theory and Practice of Security* (Butterworth-Heinemann, 2012)
10. A.S.A Statoil (ed.), The In Amenas Attack—Report of the Investigation into the terrorist attack om In Amenas. Prep. Statoil ASA's Board Dir (2013). https://www.equinor.com/en/news/archive/2013/09/12/downloads/In%20Amenas%20report.pdf

11. The Framework Regulation, regulation related to health safety and environment in the petroleum activities and certain onshore facilities (2017). https://www.ptil.no/contentassets/bf47ce8fdec745f4a2c4ed073866f079/2019_eng-26.4.2019/rammeforskriften_e.pdf
12. K. Van Nunen, M. Sas, G. Reniers, G. Vierendeels, K. Ponnet, W. Hardyns, An integrative conceptual framework for physical security culture in organisations. J. Int. Secur. Sci. **2**(1), 25–32 (2018)

Chapter 6
User Safety and Security Experience: Innovation Through Design-Inspired Methods in Airports

Ivano Bongiovanni

Abstract An inextricable organizational dilemma characterizes risk management: when effective, risk management utilizes organizational resources to avoid superior damage. When not effective, it adds costs to unmanaged risks. This clashes with growing pressures on delivery of tangible value for end-users. Safety and security management aim at mitigating risks of safety or security nature. This chapter establishes a design-based framework to re-imagine the future of safety and security in an airport security environment. The chapter proposes a method for tangible, positive end-user value delivery. Our focus is on airport security where external users live a safety and security experience.

6.1 Introduction

Despite their traditional conceptual separation within the realm of risk, safety and security management experience increasing functional connections at an organizational level. Synergies between safety and security have been explored in the literature from a systemic perspective, with the ultimate goal of better protecting Critical Infrastructures (CIs) from unintentional and intentional events, either by reducing vulnerability or by improving defenses, or both. Organizational studies that focus on the end-user of the safety and security experience are less common. In this chapter, we propose a conceptual framework that incorporates an end-user perspective to safety and security. The ultimate goal of our work is to lay the foundations for innovating safety and security experience in CIs.

I. Bongiovanni (✉)
University of Queensland, Brisbane, Australia
e-mail: i.bongiovanni@uq.edu.au

C. Bieder and K. Pettersen Gould (eds.), *The Coupling of Safety and Security*, SpringerBriefs in Safety Management,
https://doi.org/10.1007/978-3-030-47229-0_6

6.2 Background and Method

In modern organizations, safety and security are often considered two distinct sets of issues. CIs have an outstanding sensitivity to the likelihood and consequences of disruptions to their operations. In their risk management efforts, most CIs have developed unprecedented levels of granularity and specialization with regards to safety and security, which has led to the creation of two distinct organizational functions. Typical features and organizing principles have been illustrated in the literature [8], which stem from the very ontology of safety and security as defined by the absence or presence of malicious intent behind the related risks [10].

Safety and security are systemic properties and their neat distinction has been a valid approach to manage risks in times where systems were mainly composed of electromechanical constituents [16]. However, modern systems are increasingly built as a combination of sub-systems, where emergent properties are often unpredictable. The increasing relevance of the *cyber* aspects of safety (to a lesser extent) and security (the majority), besides their more traditional, physical nature, is a proof of these dynamics. Several pieces of research conclude on the need for a more holistic approach to safety and security [2, 4]), in particular disregarding the nature of systemic accidents' causes (e.g., malicious vs. accidental) and focusing on constraints that would prevent systems from being vulnerable to both safety and security accidents [16]. A holistic approach needs to be built around the commonalities existing between safety and security and various options have been proposed in the literature [9].

This chapter elaborates on some of the aforementioned features, in particular: (a) a strategic viewpoint on safety and security; (b) a mission-driven approach; (c) consideration for a broad range of stakeholders; and (d) safety and security co-design of complex, sociotechnical systems. Further, this chapter is intended to suggest an escape way from one of the most compelling dilemmas that characterize safety and security management (as components of organizational risk management regimes): processes that require extensive use of resources to prevent high-impact, low-probability events, and that do not usually produce tangible, positive value for users beyond loss prevention (Piètre-Cambacédès and Bouissou defined them as "eternal killjoys", [9]). To do so, we elaborate the concept of *safety and security experiences*, the combination of organizational policies and management and systemic features (e.g., human, technology, etc.) around safety and security, as "lived" by end-users. To configure a holistic perspective to safety and security experiences that yield value for users, we adopt a design-based perspective.

We postulate the suitability of design thinking as a methodology for innovating safety and security as a holistic user experience. Through problem-framing and re-framing, design thinking addresses problems that, like safety and security, present multiple perspectives, typical of complex systems [6]. Design thinking is a learning-driven, human-centered approach [7], suitable to shape behavioral change and human factors. It leverages the potential of collaboration among teams and stakeholder engagement to reduce individual cognitive biases [7]. Finally, design thinking synthesizes users' needs with what is technologically achievable and economically viable to yield user value [3]. We welcome Jeanne Liedtka's design thinking framework (2015) in three stages: (A) *Data gathering about user needs;* (B) *Idea generation;* and (C) *Testing* and build on research conducted in safety and security management in international airports in Australia [2].

6.3   Design-Led Innovation in Safety and Security

Design-inspired methods require to re-consider the meaning of things and look beyond what we usually do with them [15]. In airports, safety and security are traditionally organized around regulatory requirements implemented in a managerial context. One of the most common *loci* for airport security are the security screening points, where passengers are invited to surrender prohibited items and required to go through X-ray portals and have their baggage scanned. While acknowledging that safety and security have different meanings and applications based on involved stakeholders and location, we focus on the security screening points as *loci* for the safety and security experience of external users (passengers and general public). We also assume that our design-inspired methods are applied in a team setting, with participants (e.g., designers, airport managers, passenger, etc.) tasked with improving the safety and security experience.

In terms of *data gathering about safety and security users' needs* (A), a **stakeholder map** of the safety and security experience in airports would have the external users as the core of a system composed by several actors: cleaners, airline employees, retail workers, safety and security officers, etc. Such actors contribute to shaping users' safety and security experience. Acknowledgment of the variety and specificity of users' categories is one of the tenets of design thinking and a key component of building a positive safety and security experience. This is in sharp contrast with traditional airport safety and security, which tends to consider external users as one broad category, with little to no room for customization [1].

User-persona maps [13] synthesize categories of users in concise, yet deep, representations and can be created by elaborating data obtained through interviews of passengers undergoing security screening. For example, Alfred (64) is an empathetic, money-conscious and family-oriented, one-off traveler, who travels once a year to visit his daughter abroad. He has a genuine passion for learning new things. Risk-averse, Alfred loves when he and his family feel safe. He does not mind putting up with smaller "paper-cuts" (e.g., bureaucracy, time) when the ultimate goal is safety. He loves talking to people, including security officers at the screening points. He is a fan of the TV show "Airport Security" as he admires the order and firmness portrayed. On the other hand, for example, Wendy (41) is a driven businesswoman, punctual, determined, and career-oriented. Wendy has no time to waste. Single by choice, she is a marketing manager and travels for work twice a week. Smart and successful, she cannot live without her mobile phone. Wendy feels that security is an annoyance and she sometimes argues with the security officers. She cannot understand why airport security is so cumbersome, as no terrorist attacks have been recorded lately. She would spend some extra time working or shopping, rather than queuing for security.

These are two examples, and once sufficient categories of users are reviewed, **user-journey mapping** [13] is used to trace the *touchpoints* through which the safety and security experience in airports is lived. This can involve user interaction with specific products (e.g., digital technologies), people (e.g., operators), or services (e.g., floor-cleaning at the security point). User-journey maps help visualize the existing user experience to empathize with users, laying the foundations for enhancement [5]. In a user-journey map, it is important to identify the touchpoints that are most prominent to shape the user experience. Using the persona of Alfred, we can identify the following stages of a hypothetical user-journey map: 1. Research; 2. Purchase; 3. Preparation; 4. Transfer; 5. Security screening; all mapped against Alfred's feelings and actions (Table 6.1).

Alfred's journey map clearly differs from the one Wendy would have. This suggests that their safety and security experiences need to be designed differently. After review of the user-journey maps, *painpoints* can be identified and utilized as the starting point for enhancement efforts. In this stage, user interviews are essential to produce richer qualitative insights.

Table 6.1 User-journey map (Alfred)

	Research	Purchase	Preparation	Transfer	Security screening
Feeling	*"Security is a serious thing." "There is so much information, it's overwhelming"*	*"I can't wait to purchase presents for my daughter!"*	*"When I start packing my stuff early, I feel more relaxed and enjoy my trip more"*	*"If I rush things, I feel stressed and it's not a good feeling at all…"*	*"This process is so fascinating." "I am really curious about security screening"*
Doing (touchpoints)	Asks relatives and friends for a good travel agency Selects a travel agency and meets Gets all the safety, security, and customs information from the travel agent	Purchases a solid, safe suitcase Purchases presents that can be safely transported	Packs for the trip Gets all the documents printed: ticket, travel information, terrorism alerts	Allows plenty of time for security screening Reads as many signs as possible on the security process Earliest possible check-in	Reads all the indications. Prepares well ahead Talks to the operators and asks information Undergoes security

In terms of *ideas generation (B)*, to facilitate the creation of innovative ideas aimed at improving the safety and security user experience in airports, we propose using several **structured ideation lenses** [11]. For brevity, two such lenses are described in this chapter.

First, *derive* entails exploring other industries and/or businesses in search for similar problems and identify existing solutions to such problems, to derive inspiration for ideation.[1] *Derive* requires the ability to immerse real-world solutions in a context where they were not originated. Consider the following scenario: Wendy, our driven business woman, goes through airport security screening several times a week. This, as indicated in the literature [12], is a stressful experience for passengers, with potential health and safety repercussions. *How would a company like Amazon provide a stress-free experience at one of its tightest user touchpoints?* Amazon gathers as much customer information as possible to offer more personalized services. This could suggest having security and safety officers adapt their attitude toward Wendy, if they knew that she is a determined person, who does not like to waste time with security screening.

Second, *utilize* requires focusing on the under-utilized assets present in an industry, business model, or user experience and ideates original ways of leveraging their untapped potential. Consider Alfred, our one-off traveler, in his interaction with the safety and security experience at the airport security screening point; *what idle assets could be identified that airport management would want to better leverage?* The customer feedback that many airports request at the end of the screening process is an example to leverage the passenger's feelings/time.

The ultimate goal of structured ideation is to produce as many ideas as possible, and present them to all participants in the **design** exercise, to maintain a collaborative approach. However, in order to render idea generation fruitful, a limited number of final solutions has to be filtered, to check for feasibility, potential for impact, profitability, etc. [5]. This can be done in a single round or several rounds, by giving participants in the design exercise the possibility to vote for one or multiple winning ideas.

Lastly, *testing (C)* entails experimenting with the solutions deemed the most feasible and conducive of practical impact. To simplify, we assume that the participants in our design exercise ideated a **dedicated security screening mobile app** (*Sec-ScreenApp*). Wendy, the driven business woman of our example, was identified as the target user. The app could perform the following:

- Send a location-based alert when the user enters the security screening point area, to request activation;
- Provide information on the security screening process: estimated waiting time, estimated time to boarding gate, etc.;
- Offer the user information on the latest deals at the duty free, giving the opportunity to book specific products/services for later collection (e.g., a haircut, a coffee at a specific café);

---

[1] In its design sprint kit [5], Google recommends a similar framework, called 'Comparable Problem'.

- Incorporate an interactive form for user feedback before and after the screening process;
- Advise the user when they need to put the mobile phone away, as they are approaching screening.

A **storyboard** provides a representation of the steps involved with the newly designed safety and security experience and allows participants to identify the phases that need to be tested with the users and the ones that can be prototyped without testing. Figure 6.1 illustrates an example of a storyboard for *SecScreenApp*.

6.4 Discussion and Conclusion

We utilized design thinking as an approach to explore the safety and security experience of airports' external users and create conditions to innovate it. In our example, we focused specifically on security screening. *Data gathering about safety and security users' needs, ideas generation,* and *testing* are all fundamental phases in the design process. Nonetheless, helping safety and security professionals know their users better may be the most crucial priority. A user-centric approach to safety and security is not a natural perspective. Safety and security are rarely the core business of organizations. The approach we adopted meets calls in the literature for a holistic perspective on safety and security [1, 4], one that leverages their strengths to a mutual benefit. We endorse a user-centric perspective, which implies a shift from traditional legal and managerial considerations of safety and security. These different views are summarized in Table 6.2, where the *design approach* is the focus proposed in this chapter.

Building on design thinking's collaborative perspective, broad stakeholder engagement is incorporated in the approach we propose. Further, delivering a memorable safety and security experience addresses calls in the literature to overcome risk management's dilemma as an organizational function that uses resources to prevent losses, not to create tangible value [9].

It is worth mentioning that the conceptual and methodological model we propose is yet to be tested in an airport. Besides, from a generalisability standpoint, the proposed model suggests an application to airports. Its suitability needs to be tested in other CIs, but research shows how a user-centric approach can be utilized in a broad range of environments [14].

The core takeaway from this chapter is a complementary, user-centric perspective to safety and security, besides the legal/managerial stance traditionally embraced in airports. These concepts can be extended to other CIs (for example, electrical substations, train stations, or, in case of major events, stadia) to facilitate an innovative approach to safety and security, one that goes beyond loss prevention.

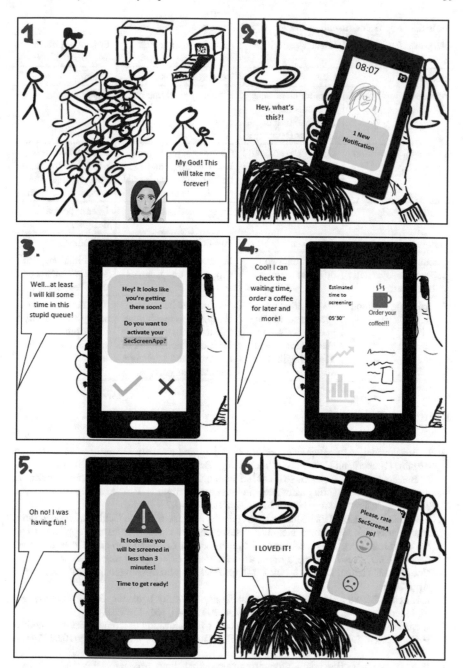

Fig. 6.1 Example of a storyboard for a dedicated security screening mobile app

Table 6.2 Approaches to safety and security in airports: a synthesis

Airport safety and security	Legal approach	Managerial approach	Design approach
Focus	Law	Resources/goals	Users
Mission	Zero unintentional accidents (safety); zero intentional incidents (security)	Efficiently mitigate risks	Delight users
Driver for innovation	Changes in societal practices cause regulations to change	Changes in regulations, business goals, and budget cause managerial practices to change	Changes in users' needs and jobs to be done change experience design
Lead-innovator	Legislator	Enlightened manager	User experience designer
Innovation source	Top-down	Top-down/bottom-up	Bottom-up
Should safety and security be separated or combined?	Separation: two different regimes	Separation/combination: depending on resources and goals	Combination: both are components of users' experience
Overarching question	"We need to meet specific standards in terms of safety and security events happening in the airport"	"We need to be compliant to regulations together with employing the most efficient mix of resources and achieving our business goals"	"We need to make safety and security a memorable experience for our users"

References

1. I. Bongiovanni, *Assessing Vulnerability to Safety and Security Disruptions in Australian Airports* (Queensland University of Technology, Brisbane, QLD (Australia), 2016)
2. I. Bongiovanni, C. Newton, Toward an epidemiology of safety and security risks: an organizational vulnerability assessment in international airports. Risk Anal. (2019)
3. T. Brown, Design thinking. Harvard Bus. Rev. **86**(6):84–92
4. S. Chockalingam, D. Hadžiosmanović, W. Pieters, A. Teixeira, P. van Gelder, Integrated safety and security risk assessment methods: a survey of key characteristics and applications. In SpringerLink (Ed.), *11th International Conference on Critical Information Infrastructures Security (CRITIS 2016), Paris, France, 10–12 October 2016* (pp. 50-62): Springer
5. Google LLC (2018). Design Sprint Kit. https://designsprintkit.withgoogle.com/. Accessed 5 April 2018
6. L.J. Leifer, M. Steinert, Dancing with ambiguity: causality behavior, design thinking, and triple-loop-learning. Inform. Know. Syst. Manag. **10**(1–4), 151–173 (2011)
7. J. Liedtka, Perspective: linking design thinking with innovation outcomes through cognitive bias reduction. J. Prod. Innov. Manag. **32**(6), 925–938 (2015). https://doi.org/10.1111/jpim.12163
8. K.A. Pettersen, T. Bjørnskau, Organizational contradictions between safety and security – Perceived challenges and ways of integrating critical infrastructure protection in civil aviation. Saf. Sci. **71**, 167–177 (2015). https://doi.org/10.1016/j.ssci.2014.04.018
9. L. Piètre-Cambacédès, M. Bouissou, Cross-fertilization between safety and security engineering. Reliab. Eng. Syst. Safe. **110**, 110–126 (2013). https://doi.org/10.1016/j.ress.2012.09.011

10. L. Piètre-Cambacédès, C. Chaudet, The SEMA referential framework: Avoiding ambiguities in the terms "security" and "safety". Int. J. Crit. Infrastruct. Prot. **3**(2), 55–66 (2010). https://doi.org/10.1016/j.ijcip.2010.06.003
11. J.C. Recker, M. Rosemann, Systemic ideation: A playbook for creating innovative ideas more consciously. 360°-the Bus. Trans. J **13**:34–45
12. S.M. Redden, How lines organize compulsory interaction, emotion management, and "emotional taxes": the implications of passenger emotion and expression in airport security lines. Manag. Commun. Quart. **27**(1), 121–149 (2013). https://doi.org/10.1177/0893318912458213
13. B. Solis, *X: The experience when business meets design* (John Wiley & Sons, Hoboken, NJ, 2015)
14. M. Tate, I. Bongiovanni, M. Kowalkiewicz, P. Townson, Managing the "Fuzzy front end" of open digital service innovation in the public sector: a methodology. Int. J. Inf. Manage. **39**, 186–198 (2018)
15. R. Verganti, *Design Driven Innovation: Changing the Rules of Competition by Radically Innovating What Things Mean* (Harvard Business Press, Boston, MASS, 2009)
16. W. Young, N. Leveson, An integrated approach to safety and security based on systems theory. Commun. ACM **57**(2), 31–35 (2014)

Chapter 7
Divergence of Safety and Security

David J. Brooks and Michael Coole

Abstract Safety and security have similar goals, to provide social wellness through risk control. Such similarity has led to views of professional convergence; however, the professions of safety and security are distinct. Distinction arises from variances in concept definition, risk drivers, body of knowledge, and professional practice. This chapter explored the professional synergies and tensions between safety and security professionals, using task-related bodies of knowledge. Findings suggest that safety and security only have commonalities at the overarching abstract level. Common knowledge does exist with categories of risk management and control; however, differences are explicit. In safety, risk management focuses on hazards management, whereas security focuses on threat mitigation. Safety theories consider health impacts and accidents, whereas security crime and crime prevention. Therefore, safety and security are diverging as distinct professions.

Keywords Threat · Safety · Professional · Body of knowledge · Concepts, practice

7.1 Introduction

Safety and security have similar goals, to provide social wellness through the management of foreseeable risks. At the abstract level, there is little to distinguish these concepts; however, at the professional knowledge level, safety and security stem from distinct basis. Distinction arises from variances in professional standing in society, task-related knowledge categories, and importantly, occupational practice. As Jore [1] suggests, safety and security frequently use the same concepts although they have separate meaning and application. Such differing views in the concepts of safety and security raise tensions across professions. To better understand and articulate the synergies and tensions between safety and security requires a better understanding of their objectives and task-related knowledge that forms and supports professional practice.

D. J. Brooks (✉) · M. Coole
Edith Cowan University, Joondalup, Australia
e-mail: d.brooks@ecu.edu.au

© The Author(s) 2020
C. Bieder and K. Pettersen Gould (eds.), *The Coupling of Safety and Security*, SpringerBriefs in Safety Management,
https://doi.org/10.1007/978-3-030-47229-0_7

63

As society becomes more complex and its members more risk averse, there will be a greater need for increased professionalism across many occupational practice areas. Safety and security are two such occupations, where both strive for professionalism. The concepts of safety and security both attempt to achieve the same goal—improving social wellness—leading to a view that there are conceptual synergies. Consequently, it is important to establish a clear understanding of both safety and security. Therefore, this chapter poses the following Research Question: *Does the body of knowledge categories of safety and security sciences demonstrate professional divergence?*

7.2   Occupational Domains

Safety is considered within the context of Occupational Health and Safety (OHS) professional practice. Security is not so bounded, given its multidimensional [2] or multifaceted nature [1]. Therefore, security is considered within the context of Corporate Security practice, being loosely defined as the provision of protection to achieve organizations goals [3].

7.3   Professionals and Their Body of Knowledge

In contemporary society, there are many emerging professions. For these professions, their development from vocational practice to a profession is challenging, specifically in social recognition. A practice domain may be defined as an area of activity or field of knowledge, over which a cultural group has occupational influence or control [4]. A cultural domain shares systems of common meaning [5] that for a profession has been articulated and codified into a body of knowledge for group consensus.

The professional has characteristics that include agreed and enforced standards of behavior, standards of education, professional development, college of peers, and a distinct and formal body of knowledge (Interim Security Professionals [6]).

A primary characteristic of a profession is its supporting academic body of knowledge. Such a body of knowledge exhibits a systematic and inclusive structure of knowledge that has logical relationships between concepts and is predictive in function [7]. Internal structure provides predictable, consistent, and reliability in the environment so that efficacy and logic prevail in professional outcomes [4]. Academic knowledge underpins and therefore, legitimizes professional work [8].

7.4 Security Body of Knowledge

The occupation of security has yet to achieve the designated status of a profession, as it lacks the characteristics of a defined body of knowledge [7]. For example, the "current body of knowledge in the security field is to a large extent fragmented and segmented" [1].

Nevertheless, educators and industrial groups [2, 4] have begun to develop a distinct body of knowledge. For instance, ASIS International has run an annual practitioner/academic symposium to develop core knowledge categories. Their outcome has been directed at United States universities in developing tertiary courses (ASIS International, 2003). In 2009, ASIS International developed a security body of knowledge (Table 7.1) with 18 knowledge categories (ASIS International, 2009, p. 44).

Brooks [2] put forward 13 knowledge categories to define security, divided into core and supporting knowledge categories. Core knowledge included security risk management, business continuity and response, physical security, security technology, personnel security, and industrial security, whereas supporting knowledge included but was not limited to law, investigations, fire life safety and safety. The study had extracted these knowledge categories from a critique of 104 international tertiary security courses from Australia, South Africa, United Kingdom, and United States. These knowledge categories were integrated and formed into a security framework [7]. The framework considers the breadth of security, whereas traditional security knowledge has generally focused on electronic, manpower, and physical security. In contrast, more mature professions selectively draw from related disciplines to define their specialization [9].

Another study linked ASIS International with academia to produce a tiered approach to security. Extending from Brooks, the Enterprise Security Industry

Table 7.1 ASIS international symposium security model

ASIS international security model		
Physical security	Personnel security	Information security systems
Investigations	Loss prevention	Risk management
Legal aspects	Emergency planning	Fire protection
Crisis management	Disaster management	Counterterrorism
Intelligence	Executive protection	Violence in the workplace
Crime prevention	CPTED	Architecture and engineering

(ASIS International, 2009, p. 44)

Table 7.2 Security professionals tasks, knowledge areas, and learning objectives

Professional task	Knowledge categories	Learning objectives
Diagnosis	Concept of security, law, security risk, assessments, survey	Contextualize security risk of organization
Inference	Security theories, physical, prevention, human factors, planning, and design	Comprehend and apply physical security system
Treatment	Security technology, detection, physical, delay, response, and procedural	Recommend and design protection system
Professional practice	Information, business, design, project, contract, and research skills	Employ knowledge to achieve objectives

Adjusted from [20]

Model [10] used a five-tier model with tier-four being industry-wide technical categories that included risk, personnel security, physical, cyber, investigations and crisis management.

A recent study [4] investigated security knowledge using a cultural domain analysis to develop physical security knowledge. As Coole et al., states, "physical security lies within the vocation of security [where] the physical security practitioner provides protective advice" (2017, p. 2). The study articulated the security professionals' knowledge areas, supported by learning objectives (Table 7.2).

These studies are not comprehensive; however, they demonstrate that there is a corporate security body of knowledge developing and that over time, a level of consensus could be gained. As Criscuoli acknowledged, security is not intuition or common sense; rather, it contains a complex body of knowledge that requires the ability to prescribe appropriate security measures for specific circumstances [11], p. 99).

7.4.1 Synthesis of Corporate Security Knowledge Categories

From these security bodies of knowledge studies, a summary of the more consensual knowledge categories are tabulated (Table 7.3).

7.5 Safety Body of Knowledge

As with security, safety has yet to achieve designated professional status that has a robust supporting academic discipline. As SIA states "health and safety is still an emerging profession that has not historically been well defined, locally or globally"

Table 7.3 Corporate security knowledge categories

Corporate security categories	
Knowledge	Descriptor
Threats and risks	Legislation and regulations; Causation and intent in crime; Security and criminology theories, models and strategies in crime prevention; Risk and security risk management; Human factors
Threats and risks controls	Diagnose, infer, and treatment controls; Physical security controls; Security technology controls; Personnel security controls; Cybersecurity and information controls; Workplace assessment, surveys, and audits; Workplace design and planning; Business Continuity Management, in incident, crisis, emergency, and recovery response
Security management	Security management, organizational culture, and societal context; Threat and risk assessment; Decision-making in risk; Monitoring, evaluating, and validating controls; Policy and procedures; Specific industrial risks, controls, and regulations; Governance
Underlying technical and behavioral discipline	Systems, human, and technology as a biological system; Social and individual psychology; Engineering and technology
Professional practice	Security information; Communication, consultation, design, and change; Organizations, project management, contract management, strategic and operational planning, business imperatives

Adjusted from [7, 10, 20]

[12]. Consequently, safety lacks a body of knowledge, where there are "substantial variations in OHS courses provided by [Australian] universities" [13] and "poor professional boundaries across the safety profession" [13].

Within the Australian context, the two more significant bodies of knowledge works have been published by the Safety Institute of Australia Ltd (SIA) and International Network of Safety and Health Practitioner Organisation (INSHPO). The SIA presented an Occupational Health and Safety (OHS) professional educational program through the Model of OHS Practice [14]. As Pryor states, this "resulted in the development and publication of the OHS body of knowledge" [15], p. 5). The intent of the Model of OHS Practice was to gain Australian university accreditation to support the professional practice of safety (Table 7.4).

The International Network of Safety and Health Practitioner Organisation (INSHPO) developed the OHS Professional Capability Framework to provide a

Table 7.4 SIA model of OHS practice

SIA model of practice components	
Consultation and building relationships	Working in an organizational context
Gather information	Apply conceptual framework
Understand the problem/situation	Diagnose/articulate thinking
Develop options for action	Decide on options for action
Operationalize	Implement actions
Monitor implementation	Evaluate change
Evaluate professional practice	Report to key personnel

Adjusted from [14]

"consensus-based tool developed to promote a higher standard of capability for OHS professional" [16]. The framework articulates OHS professional capability, where "capability" is defined as "the applied theoretical knowledge that underpins professional practice with industry-specific knowledge" [16]. The INSHPO framework is a matrix that is divided into six knowledge categories (Table 7.5), to tabulate "underlying knowledge needed to perform those tasks" [16].

These safety studies are not comprehensive; however, they do demonstrate that there is a developing international body of knowledge that is gaining a level of consensus. Furthermore, there is a clear drive by the relevant professional safety associations to integrate tertiary education within the bodies of knowledge.

7.5.1 Synthesis of Safety Knowledge Categories

From these past safety bodies of knowledge studies, a summary of the more consensual knowledge categories are tabulated (Table 7.6).

7.6 Comparison of Safety and Security Knowledge

The synthesis of knowledge tables (Tables 7.2 and 7.6) was merged to articulate knowledge categories across the two professions. There appeared to be a distinct alignment of knowledge with risk management, controls, management, and professional practice. In contrast, there were polarities with hazards and threats, technologies, and underlying theories. Commonalities in knowledge, at a cursory level, demonstrated a degree of professional alignment. Nevertheless, when these categories are explored as an occupational task, there is limited alignment in context, scope, and practice.

Table 7.5 INSHPO OHS capability framework

INSHPO OHS categories	
Knowledge	Topic descriptor
Hazards and risks	Causation in health, psychosocial, safety, and the environment
	Hazards in process, task analysis, methods, behavior, and factors
	Risk as uncertainty, hazards, criticality, and measure
Hazards and risks controls	Control principles, process, workplace design, barriers, procedures, and administration
	Mitigation with emergency preparedness and health impacts
Safety and health management	Safety management, organizational culture, law, regulation, and societal context
	Risk assessment and decision-making in risk
	Monitoring, evaluating, and validating controls
	OHS information management, communication, consultation, and change management
Role and function	Ethics and professional practices
Technical and behavioral discipline	Systems, human as a biological system
	Social and individual psychology
	Statistics, quantitative analysis, science, and engineering
Management science	Organizations, project management, strategic and operational planning, business imperatives

Adjusted from [16]

Both professions practice risk management, using the risk management standard ISO 31000:2018. For example, SIA OHS model of practice annotates the ISO 31000 risk standard (2012, p. 10) and in security, Smith and Brooks [7] present this risk standard. Furthermore, as Jore states, in "practical security risk management, the same perspectives and risk analysis methodologies seem to be shared across the security and safety fields" (2017, p. 15). However, safety and security's approach to risk management is distinct.

Safety considers risk from the perspective of hazards, which exposes someone to injury or loss. Whereas security considers risk from threat, being the purposive intent and capability of an adversary [7]. In other words, "the objective of security is to minimize the risk of malicious acts" [17]. Furthermore, threat is a central theme within the understanding, management, and application of security risk management [18].

Underlying theories for safety focuses on workplace, and resulting health impacts and non-malicious accidents. In contrast, security focuses on crime and crime prevention, as a result of malicious threat actors.

Table 7.6 Synthesis of OHS knowledge

OHS categories	
Knowledge	Descriptor
Hazards and risks	Legislation and standards; Cause in health, safety, and environment; Models of health impacts, fatigue, accidents, and environmental harm; Risk management; Hazard analysis methods
Hazards and risks controls	Diagnose, infer, and treatment controls; Physical controls; Process and workplace controls; Procedure controls
Safety and health management	Safety management, operationalize, organizational culture, and societal context; Law and regulation; Monitor, evaluate, and validate controls; OHS information; Communication, consultation, relationship building, and change management
Role and function	Ethics, professional practices; Evaluate practice
Technical and behavioral discipline	Systems, human as a biological system; Social and individual psychology; Statistics, analysis, science, and engineering
Management science	Project management, strategic and operational planning, business imperatives

Adjusted from [14, 16]

Safety and security practice the control of identified risks through diagnoses, inference, and treatment. For both professions, control includes process, workplace design, and physical, personnel, and procedural mitigation. Nevertheless, risk control has to consider whether the perpetrator has malicious intent or is accidental. Although safety control may also consider intentionality [1], intent is a significant factor in security controls. Therefore, security controls tend to focus on physical hardening to deter and delay, with technology to detect and personnel to respond. In contrast, safety controls involve people-focused approaches, with human-error and compliance issues [19].

Knowledge in legislation and regulations of safety and security suggests commonality; however, legislations are distinct. Within Australia, legislation provides explicit regulation of workplace safety. For example, "it is the law to employ or engage a suitably qualified person to advise on issues impacting the health and safety of your employees" (WorkSafe Victoria, n.d). In contrast, security has no legislation regarding professional practice except to gain a Police license to work in parts of the industry. At times, the legislation of safety drives the need for security in the protection of people from foreseeable events.

Security is multidimensional, incorporating many and diverse occupational practice areas. For instance, security sits on a continuum from national security to community security [4]. Therefore, the practice of security is difficult to define without explicit context. In contrast, safety is more commonly known within the workplace as Occupational Health and Safety (OHS). Therefore, OHS has an explicit context, resulting in a far more easily definable body of knowledge, educational learning objectives, and university level accreditation.

7.7 The Divergence of Safety and Security

The chapter posed the question: *Does the body of knowledge categories of safety and security sciences demonstrate professional divergence?* At an abstract level, there are commonalities with the occupations of safety and security, not least the drive for social wellness. Without context, there is an argument that safety and security are similar occupational undertakings, which could, therefore, be supported by a common body of knowledge. Nevertheless, commonalities only exist at the abstract level.

From a knowledge and practice perspective, each occupation considers their goals from a unique and distinct context. For example with risk management, safety reviews risk from the context of hazards management, whereas security views risks from the context of malicious centered threats. Control of risks also indicated commonality, although the inference of control treatment across the occupations considers whether the perpetrator has malicious intent or is an unintended sequence of events (accidental).

To merge the professions of safety and security to a single practice only dilutes their understanding and boundaries. Nevertheless, the International Network of Safety and Health Practitioner Organisation suggests that the safety professional has a security function (2017). However, such function is generally, in life safety, a view which is supported by Smith and Brooks who state that "life safety systems take precedence over security requirements" (2013, p. 94).

It has been argued that the occupation of safety does not draw on security's distinct knowledge basis. Considered within professional practice of knowledge categories, it was found that there is explicit and supportable divergence of task-related knowledge. Although these occupational undertakings are distinct, from the stance of future professionalism, there are commonalities within professional practice. However, beyond generic professional capabilities, divergences stand out.

While safety considerations may drive the need for security, the achievement of security is through a distinct body of knowledge. Divergence between these two occupations will be driven through greater aversion to social risk, higher expectations of professions, and with both occupations striving for professional standing. Whether each occupation will emerge as a socially recognized profession remains to be seen; however, these factors will increase the divergence of occupational safety and corporate security.

7.8 Conclusion

The chapter explored the professional synergies and polarities between the safety and security within organizations, through the insight of professional bodies of knowledge. Specifically, security was considered within the context of Corporate Security, and safety within the context of Occupational Health and Safety (OHS).

At the abstract level, safety and security have distinct commonalities, although at practice there are explicit differences. Commonalities exist within professional practice, which are generic capabilities expected within all professions. At a cursory level, common knowledge exists with risk management, risk control, and underlying theories; however, differences are explicit. For example, safety risk focuses on hazard where drivers are accidental; whereas, security focuses on threat where drivers are malicious intent. Safety considers health impacts and non-malicious accidents, whereas security considers crime and crime prevention.

Consequently, within the occupations of safety and security, and supported through their professional bodies of knowledge, there are limited synergies in underlying theory and practice. Safety and security are two distinct professions that will further diverge as each pursues professional standing.

References

1. S.H. Jore, The conceptual and scientific demarcation of security in contrast to safety. Eur. J. Secur. Res. (2017)
2. D.J. Brooks, What is security: definition through knowledge categorization. Secur. J. **23**(3), 225–239 (2009)
3. K. Walby, R.K. Lippert, *Corporate Security in the 21st Century: Theory and Practice in International Perspective* (Palgrave Macmillan, Basingstoke, 2014)
4. D.J. Brooks, M. Coole, J. Corkill, Revealing community security within the Australian security continuum. Secur. J. **31**(1), 53–72 (2018)
5. J.P. Spradley, *The Ethnographic Interview*. New York Holt, Rinehart, and Winston (1979)
6. Interim Security Professionals Taskforce, *Advancing security professionals: A discussion paper to identify the key actions required to advance security*. Melbourne: The Australian Government Attorney General (2008)
7. C.L. Smith, D.J. Brooks, *Security Science: The Theory and Practice of Security*. Waltham: MA: Butterworth-Heinemann (2013)
8. A. Abbott, *The System of Professions: An Essay on The Division of Expert Labour* (The University of Chicago Press, Chicago, 1988)
9. L.J. Young, Criminal intelligence and research: an untapped nexus. J. Australian Inst. Profess. Intell. Off. **15**(1), 75–88 (2007)
10. University of Phoenix, *The Future of Security: The Enterprise Security Competency Model* (University of Phoenix, Elwood, AZ, 2015)
11. E.J. Criscuoli, The time has come to acknowledge security as a profession. Ann. Am. Acad. Polit. Soc. Sci. **498**(1), 98–107 (1988)
12. SIA, *OHS role definitions*. Safety Institute of Australia Limited (2017). Retrieved from https://www.sia.org.au/certification/ohs-role-definitions
13. C. Chua, *Changing Landscape for WHS Professionals/Practitioners in Australia*. Presentation at the Australian Universities Safety Association Conference 2015. Queenstown: New Zealand (2015)

14. P. Pryor, M. Capra, Foundation Science. In HaSPA (Health and Safety Professionals Alliance), *The core body of knowledge for generalist OHS professionals*. Tullamarine, VIC: Safety Institute of Australia (2012)

15. P. Pryor, Accredited OHS professional education: a step change for OHS capability. Saf. Sci. **18**, 5–12 (2015)

16. INSHPO, *The occupational health and safety professional capability framework: A global framework for practice*. International Network of Safety and Health Practitioner Organisations (INSHPO). Park Ridge: IL: International Network of Safety and Health Practitioner Organisations (2017)

17. Sandia National Laboratories, *Sandia Report SAND2013-0038: Security-by-Design Handbook* (US Department of Energy, Oak Ridge, TN, 2013)

18. D.J. Brooks, Security risk management: a psychometric map of expert knowledge structure. Risk Manag. **13**(1/2), 17–41 (2011)

19. D. Borys, D. Else, P. Pryor, N. Sawyer, Profile of an OHS professional in Australia in 2005. J. Occup. Health Saf. Australia NZ **22**(2), 175–192 (2006)

20. M. Coole, D.J. Brooks, A. Minnaar, Educating the physical security professional: developing a science based curriculum. Secur. J., 1–24 (2017)

Chapter 8
Doing Safety … and then Security: Mixing Operational Challenges—Preparing to Be Surprised

Todd R. La Porte

Abstract Demands for organizational safety and security continue to increase. With incommensurable, legitimate operational requirements, tensions are to be expected. Variations in the sources of tensions are explored along with potential informal modes of accommodation. Analytical thought experiments are proposed. Prepare for surprise.

Keywords Safety–security tensions · Thought experiment · Operational resolution · Amplified complexity · Analytical surprise

8.1 Introduction

Over the past decade, public insistence on both *safety* and *security* processes has increased even as primary expectations continue to insist on reliable *operational* or mission activities. Joining these capacities involve activities that are often difficult to integrate. What organizational design and operational puzzles arise when "safety in operation", and **then** "security from external threat" are demanded from organizations and public institutions as their core technologies grow in scale and complexity? This essay explores potential implications, sets a framework for empirical examination, and ends with injunctions for executive and key operational actors.[1]

---

[1] These views have been informed by intensive field study of large-scale technical organizations operating intrinsically hazardous systems. Each faced significant safety *and* security challenges while achieving extraordinary reliability. They include nuclear power stations, aircraft carriers, air traffic control—the central empirical settings for the High Reliability Organizations (HRO) project [1]—and, especially, regular periodic study over five years at a US DOE nuclear weapons lab. Note: Space limitations prompt an unusually cryptic, spare explication of conceptual logic and compressed examples.

T. R. La Porte (✉)
University of California, Berkeley, USA
e-mail: tlaporte@socrates.berkeley.edu

© The Author(s) 2020
C. Bieder and K. Pettersen Gould (eds.), *The Coupling of Safety and Security*, SpringerBriefs in Safety Management,
https://doi.org/10.1007/978-3-030-47229-0_8

8.2   Framing Assumptions and Orienting Questions

Safety and security functions seek to assure conditions which avoid a wide array of debilitating, potential lethal events (i.e., assuring non-events). Some are associated with internal, involuntarily conditions and behaviors (safety); others with intentional acts by external adversarial actors intent on destruction (security). Safety-related functions will be activated more frequently at lower levels of intensity than most security-related ones. These will be infrequent, usually with relatively intense activity. In either case, rapid response readiness will be prized. Since all operational environments harbor persistent, irreducible ambiguity and intrinsic hazard, there will be operational surprises and "breaches in security."

Analysts and designers should expect that safety and security assuring dynamics and cultures are sufficiently distinct as inevitably to produce legitimate, continuously overlapping, sometimes reinforcing, sometimes incommensurable skills and practices. Operational leaders should expect, at least informal, accommodations between representatives of safety units and security units to limit tensions and conflict. The resulting tensions are likely to be exacerbated when contemporary measures of efficiency are incorporated into the criteria of effectiveness.[2] Operational dynamics may well become unstable and policy responses dysfunctional.

What *analytical questions* become salient when there are vigorous public *demands to greatly improve and integrate safety and security processes* with key operational functions? Consider the following:

To what degree do *Safety/Security/*reliable *Operations* re-enforce each other; conversely, impede each other sufficiently to prompt tension and conflict? Explicate in terms of the interacting dynamics between each functional pair.

As the potential for Safety ↔ Security tensions increase, what organizational policies and practices limit-exacerbate existing *operational* dysfunctions?

To what degree do tensions vary as a function of different types of institutional *quality assuring constraints* associated with (Safety, Security, Operational) activities carried out in the relevant agency domains? To what degree do tensions vary as a function of different types of national *regulatory* patterns?

What processes of anticipating, managing, and engaging potentially dysfunctional dynamics are practiced? Under what conditions are they employed? What are the dynamics and consequences of relative budget decline?

These are demanding questions—derived from suggestive conceptual speculation, analytical hunches, and in-depth observers' experience. "Thick descriptions" in the answer are meager. Crisp analytical work has yet to be done. Considerable qualitative observational fieldwork is imperative…and extraordinarily demanding. Where to allocate scarce research resources? What follows is a kind of prospective guide for adventurous empirical observers.

---

[2]In most safety/security discussions, the continuity of operational effectiveness takes a tacit second seat to assuring safe and secure social environments. In management discourse, the reverse is the case.

8.3 Imagine a Thought Experiment

Start with a "what's it like to be there" thought experiment. The intent is to frame intuitive, conceptually informed imagining; to suggest research in a "what to know next" spirit; and to set the stage for generating *hypotheses and studies* in more formal, analytical discourse (informed by LaPorte [2, 3]).

What is it like to be centrally involved with enacting safety or security functions in a very reliably performing large public or manufacturing organization? Pick your favorite ones—ones that have been in copasetic harmony, now with tensions rising. Identify situations where—when safety and security functions are each done effectively—they **overlap and then threaten to negate each other**. Imagine organizational norms and practices mixing in ways that could prompt contradictory suites of skills and interacting episodes. How are these contradictions recognized? What conditions make it difficult to avoid them?

What (national) institutional conditions enable operators, citizens, and social leaders to "prepare to be surprised" ... "to be unprepared"? To what degree do these conditions limit the likelihood of institutional resilience in the face of serious shortfalls in social safety, in national security?

A. Initial Bearings

Take the *vantage of operators, members of the teams* that enact varying task requirements and assure effective network experiences within large-scale organizations. Locate groups that have confidently bounded the tensions intrinsic to integrating the different intensities of safety and security regimes—under the eyes of wary regulators/overseers. *How might the operators' views of such situations be framed?*

Responses hinge, in part, on the tacit and explicit functions, tasks and social structures clustered under the primary orienting **concepts**—safety, security, operations, and, in part, on the **activities** carried on, in the field, by those actors who have been assigned safety or security missions. Who think that's what they're doing.

In examining these situations, assume the following bounding expectations or situations. You can expect widely varying operational settings.

Situation 1. Most agencies operate where safety activities and security (watchfulness) functions predominantly complement each other in daily interaction. Requisite activities are modest and within reinforced, de-conflicting tolerances— many in the satisficing zone for safety/security management demands.

Situation 2. Intra-operational "Safety–Security" anxieties vary in response to the *degree of perceived external demand for increases in safety and/or security measures* for different arenas of operations.

B. Operative Assumptions

Current technical and environmental changes will continually increase the relatively hazardous nature of operations such that both increased densities of Safety and Security regimes will be demanded.

Policy demands almost never call for overt reduction in Safety or Security capacity once these have been established. "Hazard potential" and "environmental vulnerability" will change in one direction—greater relative internal hazard and increased external vulnerability.

Situation 3. Each Operational domain's technical core and action dynamics are associated with functionally well-vetted *skill—authority relations, both* intra- and inter-organization. These are carried out by enacting teams, ranging from those evincing no overlap in Operational, and Safety or Security personnel, i.e., stove piped silos, to teams where the same members carry out *all three functions* sorted out in different, fully integrated, networked arrangements. These include recognized action options and nested authority configurations.

Situation 4. Path dependence effects are determinant. Each setting is shaped temporally by *"Who got there first"* relationships where those who establish initial operational dynamics of prime importance set the stage of second-comers' experience. Thence, organizational responses to policymakers' demands for *integrating* Saf and Sec functions are shaped predominantly by *which functions—Safety or Security—were established first*. In consequence, Safety ↔ Security tensions, if they emerge, are likely to take on different manifestations depending on the establishing (Saf or Sec) sequence.

8.4 Imagining Safety–Security Interactions and Outcomes

Now, within the context of these expectations, (to be verified in the field), take on an **experimenter/observer** role in exploring the behavioral dynamics likely to emerge following insistent demands for the integration of both safety and security regimens. As each condition noted in Table 8.1 below is considered analytically, imagine the effects on *operators' and mid-level managers'* daily network dynamics of *established* **Operational** *regimes*. Each varies or shapes operators' experiences as they enact the technical requirements for Safety and/or Security missions. What behaviors could be an observer's targets for attention?

First, what are the salient skills and experiences needed to assure smooth, effective responses in the domains you usually engage? Second, what reactions are likely

Table 8.1 Conditions shaping operators' experience	1. Robustness of Saf/Sec conditions (V1) and policy demands (V2)
	2. Layering of Agency Safety-Security enacting measures
	3. Relative operational scales reflected in Safety, Security regimes
	4. Public's expectation (and tacit understanding)

Note Due to space limitations, **Conditions 1 and 2** are given the most attention

Security regimes

	Well in place (Hi)	Needs improvement (Lo)
Well in place (Hi)	Nuclear power ops Navy carriers I	NASA II Air Traffic Control Electricity grid
Needs improvement (Lo)	Intelligence IT III Military units Weapons labs	Hospitals IV Local/regional gov.

Safety regimes (row axis label)

Fig. 8.1 Safety/Security states when changes are demanded

were new *Safety* or *Security* changes to be demanded by political overseers and/or regulators?

Condition 1. Variations in the operating conditions (V1) and potential operational changes implied by and concurrent with increased policy demands for Saf/Sec integration (V2). An optimum initial state would be one in which (i) Safety and (ii) Security capabilities are seen as **fully established and satisfactorily in place** at expected hazard levels (i.e., "all is well", whatever the level of expected capacities). But situations vary and demands for improvement can escalate, requiring that **Saf or Sec regimes intensify and become fully integrated.** Such changes in policy insistence can dramatically affect the experiences and interactions of operators with middle management, and these changes can upset existing operational equilibria. Imagine a study sample that could inform these variations (see Fig. 8.1 for US examples).

Cell I—[Hi/Hi] **Security** and **Safety** both fully established and sustained. This is a rare combination—internal hazards and threats to systems are believed to be considerable and harbor potential for tactical/strategic attack, therefore, elevated contingency. E.g., Nuclear aircraft carriers and nuclear-powered electricity production. Key question: What happens when sustaining resources are overwhelmed?

Cell II—[Hi/Lo] **Ops, Safety well established**, say, in civilian agencies, that come to face rising hostility and threat, with subsequent demands to "ramp up" **Security** capability. E.g., NASA, aviation, transport. Key Question: Are established Safety groups allowed flexibility to accommodate and remain effective? Are their resources diverted to Security functions?

Cell III—[Lo/Hi] **Security well established**, say, quasi or fully military agencies, become pressed to reduce injuries and internally caused facilities' damage with ramped up **Safety** systems. E.g., US Weapons labs, formerly carriers, some intelligence services, and local police forces. Key question: Can security group operations complement safety newcomers?

Cell IV—[Lo/Lo] **Both Safety** and **Security** need increased operational capability. Organization in deep hole—history of injury, and damage to neighborhood *and* now facing hostile actions from groups close-in, i.e., under duress. In the past, limited need, tacit public expectation for all public institutions—due to limited external threat—and perceived limited internal hazardous context (save external fire, weather, earthquake—non-human threat). E.g., NASA; NOAA academic/analytical organizations. Key question: To what degree do leaders appreciate the steep increase of resource likely to be implied by public demands?

Analytical Challenge: **Think of circumstances that exhibit these variations. Based on whatever conceptions of organizational dynamics you favor, predict several highly likely resolutions to the tensions that you have conjured**.

Now, two central analytical questions: Given the analytical basis for your predictions, how straightforwardly were you able to call out dynamics that could be sought in field observation? Where are the analytical shortfalls, if at all?

Other operational contexts with different, more detailed levels of safety or security activities associated are likely. This suggests the second of the several shaping conditions noted above.

Condition 2. Coping with escalating hazards and threats. As intrinsic hazards and external vulnerabilities increase, we are likely to witness a series of *layered, increasingly stringent, and militant operational responses* to demands for integrated Saf and Sec capacities. These unfold as a function of increasing internal hazards and escalating external threat. Additional safety and security functions are levied within and **on top of** increased technical training and skill imperatives needed to enact core technologies and infrastructures. These additions vary from local, on-site safety preparedness programs to high alert, whole system protection from external attack. Schematic Fig. 8.2 indicates some examples of (i) increasingly demanding

Safety First Priority: *Central Challenges*

... Work place -- Safety only – interior and on-site transport

   ... Local hazmat (chem-radiation prep and inventory)

      ... Firefighting capability   *Challenges of operating reliably*

       ... Drug watchfulness (testing...)   *for a number of work generations.*

      ------------- shifting to Security ----------------------

      ... IT firewalls   *To enhance trustworthiness in face of*

... Physical intrusion –   *system **surprises** and clever adversaries*

   ... Imported hazard chem-bio   ***assuring (some) unpreparedness.***

      ... Personnel breach- secure areas; boundary control 24-7 surveillance

       ... Attack teams   1st respondents, intel and counter-intel capacity.

Security First Priority:

   (readers will know other operational challenges within this range of priorities)

Fig. 8.2 Escalating challenges. Layered operational responses

hazards—from workplace to organization-wide risk responses (emphasizing "safety first"), and (ii) external security-evoking threats (and security priorities).

The specific safety hazards and security threats are evident in a number of modern organizations both public and private. Responses to them also amplify several central management challenges noted in the Figure. All of these items were evident in the operation of large nuclear weapons facilities. The reader will likely know of other hazardous/threatened settings that have introduced other forms/layers of response ... capacities and behaviors, by and large, invisible to most institutional observers. The phenomena are worthy of careful qualitative description.

Two additional conditions also shape operator experience and challenge managerial wisdom. These should be integrated into any serious field study.[3] Analytical readers will see how the patterns above could be applied here as well.

Condition 3. Variations in relative social scale. The greater the social scale, the more complex, differentiated, and interactive/interdependent the organization **and** the more likely the emergence of triply nested—ops., saf., and sec.—authority patterns and latent resistance networks. We expect that when Operations are massive/highly complex, Safety is more diffuse, a moderate fraction of the whole, while Security is relatively limited, in the shadows (ready to emerge!).

Different operational steady states vary from Saf stable, limited regimes (with Sec mostly latent) to full integration of Saf and Sec regimes. Differences are evident in these brief examples:

Education: Small Saf, Sec tiny.
Air Traffic Con: Saf clear presence, Sec modest, latent ready to assume command.
Aircraft Carriers: Saf modest, Sec modest, and lurking.
Weapons Lab: Both Saf and Sec evident and fully manifest continually.

Condition 4. Variations in public expectation. Organizations attempting to increase the integration of security and safety systems do so in the context of the public's understanding of and insistence on effective safety/security programs and their usual unwillingness to fund these developments. These vary widely from (a) limited experience, high expectation with reluctance to fund (Lo), to (b) clear recognition of risks and willingness to carry the costs with some forgiveness and tolerance for the struggle involved (HI). "Best cases" are rare and include our experience with nuclear power plants, national weapons labs, and submarine operations.

---

[3]Editorial constraints result in only brief mention. Explication awaits more fulsome possibilities.

8.5 Amplified Complexity and Operational Switching Regimes

Thus far we have argued that institutional policy and technical developments persistently result in **additional hazard-increasing functions** (to be safely done), many accompanied by the **increasing potential/costs of massive damage** (at physical/psychological scale). In consequence, public and overseer demands intensify for both safety- and security-enhancing measures with the explicit expectations of effectively negating untoward events and hazard-prompted damage.

These efforts are launched into operational domains already exhibiting extraordinary variety in (i) their established (entry) conditions; (ii) the range of emergent counter-threats; (iii) the heterogeneity of security response measures; and (iv) an array of operator-overseers demands and dynamics with considerable tension/incommensurable potential. These phenomena are quite variegated and analytical or empirical explication is relatively sparse.

There are likely to be patterns that have escaped, even attentive observers. An especially interesting one wants exploring—patterns whose subtlety and presence are likely to be overlooked without a relatively careful field study.

This is related to the first *central analytical and management challenge* posed at the outset: "Identify operational situations in which (Saf/Sec) **functional overlaps— when done effectively—negate each other**."[4] These are organizational maelstroms where norms and practices are mixed in ways that evoke contradictory suites of skills and interacting episodes. They confront operating teams and managers in niches of high tension and can be loci of unexpected modes of coping.

To put a sharper point on this, the piling up of well-performing safety and security augmenting functions and teams also comes with (requires) coordinating and regulatory assuring networks and personnel grafted on to or embedded in the existing operational community. In effect, social complexities are amplified in ways that often distort former relationships and harbor high "apraxic" potential [2]. This presents persistent challenges for operating personnel to evolve "on the ground" accommodations that reduce the destructive potential of intrinsically incommensurable activities.

Organizational skill groups exhibit dual or parallel Saf/Sec competences, each with potentially different operational rules. This situation could precipitate *planned and practiced regimes shifting* from (a) predominantly Saf processes to Sec ones, or from (b) predominantly Sec regimes to Saf emphases when conditions about which there is "high consensus" emerge.

---

[4]Posed here in a "what's it like" thought experiment mode, it was initially framed as "To what degree do *safety/security/* reliable *operations* re-enforce each other; conversely, impede each other sufficiently to prompt tension and conflict?".

There *are* persistent cases that demonstrate high consensus about conditions for *shifting from safety to security privileging conditions*, i.e., switching **from Safety-assuring to Security-enhancing** priorities. Tipping points are recognized by all operational and overseeing institutional/legal entities, thus smoothing the organizational grounds for actualizing operational changes.

Imagine what could be expected from operators and managers in the face of such dissonant potential? An Assumption and two Hypotheses.

Assumption: Due to increasingly heterogeneous hazardous operations across widely variegated service geographies, there is a growing range of different situations *known to experienced on-site operators* but beyond the knowledge capacity of most superordinate managers unless they have had direct, "close-to-hazards" operational experience in the relevant operational domains.

Hypothesis 1: Operators and "close to hazards" managers, in efforts to limit confusion in the face of rapidly unfolding safety- or security-threatening situations, develop high-consensus rules of engagement and operational activities about shifting from one to the other processes. Each set has activities and triggers that do not reinforce those of the other.

Hypothesis 2: "Switching" protocols from one set of priorities and procedures to the other activity are mainly the province of "close to hazard" working groups and supervisory management.

Managerial (and research) imperatives follow. As patterns of unfamiliar, risky situations and/or novel threats intensify, it increases the imperatives for senior leadership to (i) insist on the **adoption of new skills**, (ii) legitimate **the capacity for switching internal emphasis** (via facilitating personnel and team development and training); (iii) assure continued **support from overseers for such dual capacities;** and (iv) maintain/enhance public understanding and **forbearance for respective needs**.

At the same time, the conditions that produce the need to (i) intensify system safety, i.e., greatly reducing the experience of known hazards and (ii) fend off aggressive external attacks meant to cripple or destroy (i.e., assuring the absence of predatory suffering) **also increase** the operating social complexities **beyond careful comprehension**.

While deep knowledge limits some surprise, internal scale and increasing complexity guarantee it. And hostile external efforts at deception and seeking destructive advantage brook persistent, inevitable unpreparedness.

What if experiencing surprise and unpreparedness were to be expected, and experienced without blame and with some sympathy for those close to the hazard, what patterns could (perhaps should) become evident from analysts and leaders?

8.6 Preparing to Be (Legitimately) Surprised... to Be (Legitimately) Unprepared

An important **emerging analytical challenge** would (and should) be the development of credible skills, norms, and practices associated with "preparing to be surprised" during the deployment of measures expected to improve safety and/or security [3]. Surprises associated with each of these domains are likely to be systematically different. This a function of distinctive: (a) sources of relative hazard and technically/socially induced vulnerabilities and system complexities and (b) societal variations in public awareness/acceptance of surprises (forecasting incompleteness) and/or institutional unpreparedness, esp. in face of opponents' aggressiveness and their success in finding weaknesses and vulnerabilities.

What overriding responsibilities and obligations for **mission and institutional leaders** would follow?

Capable senior leadership would facilitate (a) the **evolution of wary trustworthiness and empathy** for veteran safety and security first responders and stewards and (b) the emergence of **institutional cultures according to high salience to** *forbearance and forgiveness* as well as learning and blame-putting.

These would entail leadership **obligation**s to

Resist calls for "efficient" spare-ness—the thinning of watchfulness with the resulting increase in social apprehension—and to explicate/highlighting the continuing dilemmas associated with short-term impatience and political tendency to under-resource conditions of watchfulness needed by subsequent work generations.

Assure organizational and public understanding of Saf/Sec stewardship roles and their fundamental contributions.

Enhance a sense of honor and resources—beyond operational costs—for safety and security stewards. And remind us of our underlining dependence on those who, in effect, have signed up **to take a bullet, suffer severe burns or serious injury on our behalf**.

Afterword Insisting on enhanced operations of both safety and security functions across the critical organizations and institutions is increasingly likely and imperative. Responding to these imperatives—the ken of institutional leaders, senior managers, and, especially, experienced supervisory veterans—is already challenging organizations across a wide spectrum of social life. Those responsible take up the tasks often with little preparation and limited experience. Making rapid progress in understanding the conditions of both their copacetic joining and potentials for disabling dysfunctions, especially, the sources of intra-operational conflict is essential. Seeking these improvements becomes a major challenge for us—the community of analytically acute observers of organizational life.

This essay touches on only some of the more obvious variations in organizational situations that will shape different outcomes. Reviewing the chapters of this book reveals something of the wider sweep of significant factors and the breadth of the challenges of knowing. We are still just a little beyond the starting line. A detailed sense of how these factors shape the outcomes and of the behavioral consequences awaits a significant increase in empirical acquaintance and rigorous fieldwork—often of the most demanding sort. It is also a work that is likely to be as intriguing and interesting as it is important.

References

1. T.R. LaPorte, High reliability organizations: unlikely, demanding and at risk. J. Crisis Conting. Manag. **4**, 2, 60–71 (1996). (Special Issue on New Directions in Reliable Organization Research, ed. G.I. Rochlin.)
2. T.R. LaPorte, Observing amplified socio-technical complexity: challenges for technology assessment (TA) regarding energy transitions, ch 12, in *Energy as a Sociotechnical Problem: An Interdisciplinary Perspective on Control, Change and Action in Energy Transitions*, ed. by C. Buscher, J. Schippl, P. Sumpf, Routledge Studies in Energy Transitions, (2018), pp. 245–261
3. T.R. LaPorte, Preparing for anomalies, revealing the invisible: public organizational puzzles. Risk Hazards Crisis Public Policy **9/4** (2018). https://www.researchgate.net/publication/324 111441

Chapter 9
Safety and Security: Managerial Tensions and Synergies

Paul R. Schulman

Abstract The relationship between organizational safety and security is a conceptual and practical challenge. This paper focuses on the management aspects of this challenge. Its argument is that we have yet to parse out the full range of contradictory and complementary requirements of these two as managerial missions. Considering the requirements for high reliability management can provide a clarifying lens for sorting out the contradictions and complementarities. Some overlapping requirements from a high reliability perspective actually argue for an integration of the two missions within one managerial framework with enhancements for "higher resolution" reliability.

Keywords High reliability management · Safety management · Security management · Vulnerability

At a recent conference on safety management organized by a large public utility regulatory agency, the issue of infrastructure security came up for discussion. Addressing the issue, the CEO of a large utility asserted: "If we're doing a good job of safety management that should take care of security too". I took this to be a convenient untruth at the time.

There is a strong debate among scholars and practitioners about whether the same managerial framework within a single organization can accommodate both effective and successful safety and security management [1, 3, 5]. Research on High Reliability Organizations (HROs) has focused on a number of organizations (nuclear power plants, commercial aviation and air traffic control centers, and electrical grid management organizations, for example) with extremely well-developed reliability strategies in both technical design and management systems for protection against failures that can create catastrophic accidents [6, 7, 10, 11]. These organizations notably, as critical infrastructures, are also potentially high-value targets for terrorist assault. But it is not clear from this research that HROs are simultaneously addressing both safety and terrorist security objectives in their reliability strategies.

P. R. Schulman (✉)
Mills College and University of California at Berkeley, Berkeley, USA
e-mail: paul@mills.edu

C. Bieder and K. Pettersen Gould (eds.), *The Coupling of Safety and Security*, SpringerBriefs in Safety Management,
https://doi.org/10.1007/978-3-030-47229-0_9

This essay will consider and compare major variables that have to be addressed and the strategy developed by an organization seeking high reliability management first, with a safety and then, a security mission.

9.1 Safety Variables and Strategy

The most prominent feature of HROs is that they are managing technical systems that can fail with catastrophic results—large-scale disruptions of critical services and potentially many deaths. "High Reliability" for an HRO means that there are protections against these failures or accidents in place to preclude them from happening—not just probabilistically, but deterministically. A key managerial feature of this reliability is the protection against errors that could lead to these "precluded events", especially "representational errors"—mis-estimations, mis-specifications, and misunderstanding of the systems being managed that can lead to decisions and actions that invite failure and accidents. In this sense, reliability strategy is simultaneously a commitment to safety because you cannot ensure safety without reliability. But it is system safety, not individualized accidents such as slips, trips, and falls, that is the priority of HROs.

Robustness of technical systems. The technical systems under management are well understood in terms of operating principles and the maturity of technical designs. These are not frontier technologies whose operation is experimental in both underlying knowledge and operational experience. Much if not all of operations and maintenance is carefully analyzed, including careful risk analysis, and conducted under elaborate procedures. In the United States, for example, it is against federal law to operate a nuclear power plant "outside of analysis".

Robustness is supported by technical designs that include redundancy of key components, back-up systems to compensate for the loss of primary ones, as well as planned and even automated shut-down protocols to stop operations in safe modes relative to potential major accidents. Non-operation is always a priority to continuing to provide outputs in the face of escalated risk.

The reliability of technical system components is often defined as how well their designs fulfill operational requirements and performance expectations, and within these designs how infrequently they fail. But reliability cannot be fully determined by designs alone. Components must be inspected, operated, and maintained within design specifications. This requires their protection by management from errors that could undermine these processes. Assuring the integrity of management information, decision, and control processes to prevent error is a major feature of managerial reliability.

HRO managerial strategy. A classical HRO management strategy for reliability and safety in managing technical systems is founded in the formula that low *input* variance (in external resources, supports, demands, and conditions surrounding the organization) coupled with low *process variance* (operations tied to procedures and careful prior analysis and planning) lead to low output variance (predictable and reliable performance). Control over input and process variance are key elements in stabilizing performance.

Yet, ironically, this control is grounded in the recognition that a key to high reliability is not a rigid invariance in technical, managerial, and organizational processes but rather the *management of fluctuations* in task performance and conditions to keep them within acceptable bandwidths and outside of dangerous or unstudied conditions [12]. Many organizational processes that support high reliability, including high degrees of attention and care in specific tasks, lateral inter-departmental communication and trust, and shared sensemaking surrounding the execution of plans and decisions, are perishable in the press of day-to-day work and have to be continually monitored and renewed.

Supporting this narrow bandwidth management is the careful identification analysis, and exclusion of precursor conditions that could lead to precluded events. HROs begin with the core set of these unacceptable events, then analyze backward to conditions both physical and organizational that could, along given chains of causation, lead ultimately to significant possibilities of such events. This "precursor zone" typically grows outward to include additional precursor conditions based on more careful analysis and experience. These precursors are in effect leading indicators of potential failures and are given careful attention by operators, supervisors, and higher managers.

Some precursors are in effect "weak signals" to which "receptors" throughout many levels of the organization are attuned and sensitive. Examples of precursors observed in HRO research included: operating equipment nearing the edge of maximum allowable conditions such as temperatures or pressures; too much noise or too many people in a control room; silence or edginess in an individual control operator; backlogs in clearing corrective action reports; a movement into "unstudied conditions" in any operations or maintenance activities. In its effectiveness, this process of precursor management provides a special kind of "precursor resilience" to these organizations. They can identify and move quickly back from the approach to precursor zones while still maintaining a robustness in performance and outputs [10].

Another important element in HRO reliability management is the existence of a great deal of lateral communication. This is important to maintain the system focus of reliability and safety management and prevent localized actions without consideration of their wider effects. There is a lot of inter-departmental collaboration in work planning sessions, incident investigations, procedural reviews and procedure revisions.

Additionally, there is a widely shared culture throughout these organizations that supports the features described above. This culture supports managing to worst-case possibilities and not simply probability, and in many decision-making and planning activities. There is highvalue and indeed much personal esteem accorded to individuals who can offer imaginative examples of potential causal pathways to worst-case possibilities. The culture within HROs also stresses widely dispersed individualized responsibilities associated with detecting error, such as speaking up to correct it, and promoting the identification of precursors and the improvement of procedures. In one HRO, many individuals down to the control room and maintenance shop levels, for example, actively participated in the procedural revision process. In important respects, these individuals "own" the procedures. In one nuclear power plant personnel at different levels expressed the view that, without continual improvement, existing levels of reliability would likely not be maintained due to the onset of complacency [12].

One important example of the culture of responsibility for safety reaching down to the level of the individuals is the importance of people we termed "reliability professionals" to the successful pursuit of reliability and system safety in HROs [11]. Who are reliability professionals?

These are individuals who have special perspectives on reliability, cognitively and normatively. They mix formal deductive knowledge and experiential knowledge in their understanding of the systems they operate and manage. Their view of the "system" is larger than their formal roles, specializations and job descriptions. They internalize in their identity the reliable and safe operation of their systems. In this, they are "professionals" on behalf of reliability and safety, but are not defined by particular formal degrees or certifications.

We have found reliability professionals distributed in many jobs at many levels in HROs. We find them among control operators, line production or maintenance personnel, engineering and other technical personnel who support operators and maintenance, and among middle-level managers and supervisors, department heads, and CEOs or agency heads. Whatever their formal job, they focus on identifying precursor conditions that degrade safety, including their own performance capabilities. They can also help police their own departmental or unit movements toward a practical drift away from reliability and safety because in their larger system perspective they think about the system risks and consequences of changes they or others make in their own task domain [8].

All of these elements in reliability and safety management in HROs reflect the widely stressed idea that, despite all the prior anticipation and analysis, the elaboration of procedures, the redundancies, and shut-down protections, there is still the potential for surprises in their technical systems and a constant need for vigilance and organizational improvement.

But it is important to appreciate that in this recognition of the *potential* for surprises and their strategy of precursor resilience, HROs are hardly confronted with major uncertainty in day-to-day operations and performance. In their settled technology, elaborate planning, anticipation, and analysis, it is not "managing the unexpected" HROs are engaging in. Instead, in managing to possibility, and adding a worst-case slant to planning and analysis, they are *enlarging* expectancies—formalizing an approach to avoid complacency and add to the possible scenarios that are part of their prior analysis and anticipation.

Now let's consider management challenges with respect to a *security* mission and what "high reliability" might mean in this context.

9.2 Security Management Variables and Reliability Strategy

Failure versus Vulnerability. One obvious difference between safety and security management is in the primacy of hostile intent. For example, while they may have been "hardened" to resist an external assault, and while their management systems were well organized to guard against unintended errors in operations and maintenance, HROs were not well prepared to protect against willful and strategic *internal* sabotage through actions of destructive intent.

This is a special challenge in "managing the unexpected". It is one thing to identify and manage operational risks of failure, it is another to identify and manage vulnerability to destructive intent. There are always more ways that a complex system can fail than there are for it to operate correctly as designed. But hostile strategy, both external and internal, can add additional possibilities for disaster because of the treatment of vulnerabilities as strategic targets. Further, if attacks on these vulnerabilities do not have to include the survival of the attackers, the possibilities get even larger still. An example of this is the strategy adopted by airlines after 9/11 to harden the cockpit door of airplanes to resist external intrusion from potential terrorists among the passengers. Ironically and tragically, addressing this problem led to a reciprocal vulnerability: a saboteur already inside the cockpit. In fact, a suicidal co-pilot of Germanwings Flight 9525, with the pilot momentarily out of the cockpit, locked the door, thus making himself impregnable, and flew the plane into a mountain-side. Protecting the cockpit against external intrusion actually created a new vulnerability and an opportunity for a different form of assault. Achieving reliability and safety against nature or inadvertent human actions is hard enough. It becomes a different challenge when failure itself is part of a learning system with the ability to develop *counter-strategy* for its promotion.

A special problem in "design-based" vulnerability. Vulnerability itself can come in a variety of forms, giving many options to saboteurs. Vulnerability means risk exposure, but vulnerability also pertains to an *innate ability to be harmed*. One form of vulnerability is by willful design strategy plus by potential victims toward harm itself. In way, victims move to make an assault more likely and/or more consequential with respect to harm. An example is when housing developments are built in flood plains, increasing their vulnerability to floods or when tall buildings or roads are constructed on earthquake faults or individuals build homes in forest areas with increased exposure to wildfire. A spectacular example of design-based vulnerability lies in the internet and its vulnerability to cyberattack.

It has been argued that the internet can now be attacked from any location, at any scale, and across a wide range of precision [4]. No natural system on earth could survive to evolve such an extreme degree of vulnerability. But the internet is not a natural system. We have allowed, encouraged, and designed it to evolves to this high degree of vulnerability. Currently, the internet is certainly one of the most important critical infrastructures with simultaneously the most extensive social dependency and the highest vulnerability of any system humans have ever created.

Every new element we add to internet connectivity, or the extension of its functions and capacity, introduces additional vulnerabilities—often across multiple dimensions. This design perversity, in which each new design element adds an increased number of vulnerabilities, to viruses, hacking or fraud, is an enlarging challenge to our processes of forecasting and understanding. It is hard to see how, under these current challenges, internet security can be successfully managed by individual organizational strategy or effectively addressed in local or regional public policy. By extension, it would seem that controlling design-based vulnerability implies the need for larger-scale social regulation than organizational self-interest or even an industry-wide self regulation would fully address.

In addition, for a terrorist, not every target has to be of high value in terms of disruption and death. Terror is designed to induce public fear and uncertainty, as well as policy reactions in anger that may lead to the sacrifice of other values held by a society. In this way, targets can have symbolic value well beyond any physical destruction. Even attacks on targets of marginal significance, or attacks that fail, can induce fear and a sense of vulnerability within a population. Security management can hardly be the management of everything. As two analysts conclude, "Most sensible people would […] agree that it is impossible to thwart each conspiracy and detect each and every lone individual or group harbouring evil intentions" [2]. So high reliability security management cannot realistically rise to the level of the precluded event standard sought for safety in HROs, nor is it likely to be pursued effectively by single organizations.

Managerial control variables. Strategic vulnerability adds additional challenges to reliable security management. Targets can be both external and internal. While an organization may have a set of controls it can use in internal operations (hierarchical authority, procedural requirements, training, hiring and firing, surveillance, etc.), it may have few control variables to cover the vulnerability of external infrastructures, or the loss of goods or service outputs from other organizations upon which it depends for operation. Further, those attacks that are not prevented may well require coordinated emergency response and recovery operations under conditions difficult for any single organization or set of organizations to manage effectively:

> Plans for crisis and disaster management tend to have a highly symbolic character, and often provide little guidance for those who must respond to unforeseen and unimagined events. In addition, plans are only useful if they are tested and refined and the people who work within the plan are familiar with their roles, responsibilities and interactions with others [2].

Also, consider the risks of risk assessments themselves applied to vulnerability. The purpose of risk analysis and assessment is to identify risks, rank them in terms of their importance and likelihood, and prioritize attention and resources devoted to them on the basis of this priority. Yet security vulnerability assessments, if known, might well undermine their own accuracy through counter strategy they might generate on the part of hostile strategists. They could in fact attract potential terrorists and add to the risk surrounding what are identified as vulnerable targets, or instead increase risk to low risk -assessed targets. Why not direct hostile action to what appear to be the *least* likely and possibly least defended targets?

The role of precursor management and the search for leading indicators and weak signals may also be limited in security management because perpetrators of attacks will make every effort to keep potential precursor actions and information secret or even disguise their intentions with false signals.

Contradictions between security and safety management strategy. Given the challenges associated with security management, it is not surprising that some of the approaches taken toward managing vulnerability might conflict with those dominant in safety management. The stress on anticipation and prior analysis appropriate to reliable safety management of settled technical systems may introduce rigidity and undermine the resilience and adaptability necessary for rapid responses to unexpected assaults. The need to contain the spread of information, about key plans, decisions, and priorities, within an organization to prevent counter learning on the part of internal or external enemies, may conflict with the system-level perspective, supported by extensive lateral communication, relied upon by HROs. Even the effort to harden targets against attack through guns, gates, and guards, as in the Germanwings example, may cause security protocols to conflict with the ease of access necessary for collaborative operations and decision-making [9]. The frameworks for physical and information security may also interfere with rapid inter-organizational coordination of emergency response operations after an attack.

9.3 A High Reliability Perspective on Both Safety and Security

A case for overlapping management. Even given the differences and potential contradictions between safety and security management, it could still be argued that there can be constructive overlap or "synergies" in the effort to achieve high reliability management of both missions. As noted in the workshop prospectus, failures in either can lead to "similar ultimate consequences" which may require similar emergency and crisis management approaches (although in terror attacks, first responders may also be part of the intended targets).

A major foundation for an overlap of high reliability management of both safety and security is the common need in both missions to identify and minimize error. Managerial reliability is founded on the management of error—including the errors of misperception, misidentification, and misunderstanding. The constant search for these forms of representational error, the continual questioning of assumptions, and accepting the possibility of surprise are underlying strategic and cultural features of high reliability management for safety. They are useful in warding off complacency and hubris, states of mind that could undermine the flexibility and imagination useful in foreseeing and preparing for terror attacks. The pattern recognition skills of reliability professionals are likewise useful for identifying quickly both unfolding system failures and progressing terrorist assaults.

The focus on precursor conditions and precursor indicators is also important to both missions. Learning to recognize potential precursors to terror attacks is already an important part of security management. Here again, the support given to reliability professionals in high reliability management is important. These individuals are often ones who look for and detect weak signals. In organizations with high reliability management, there is generally a receptivity to spokespersons for a neglected perspective. Terms such as "I find myself in uncharted territory" or "I'm not comfortable taking this action" are taken seriously, especially when those using them also have the ability to stop work or veto actions as part of their job authority.

Challenges in safety and security management integration. It will not be easy to bring both safety and security missions under a larger framework of high reliability management. This will require developing both a wider scope and longer time frame of anticipation and at the same time, a more distributed participation in error detection and precursor resilience. A higher resolution anticipation entails gathering a wider range of information and applying more imagination to the analysis of both external and internal threats over a range of scales, scope, and time.

Enlarged time horizons for anticipation and planning could uncover both slow motion safety and vulnerability issues, such as the design-based vulnerability described earlier, or even climate change, that are likely to grow over time. Enhanced anticipation will likely need to be matched by a wider range in the *scale* of reliability management to include recognizing inter-organizational and even international precursors of accidents or assaults and the ability to manage defenses and responses on these international scales.

Teams also can be important to enlarging the range of organizational *resilience* in the face of failure or attack. In many cases, reliability professionals in teams function effectively as first responders after failure or attack, bringing their experience and effectiveness at pattern recognition quickly to bear in guiding action to limit damage or speed recovery. The role of U.S. air traffic controllers in clearing the skies of all aircraft quickly after the onset of the 9/11 terror attacks is a good example of this emergency response role at the service of both safety and security. Organizations can train and support their reliability professionals to use their skills as a means to improve both the anticipation of vulnerability and the capacity for resilience to back-up reliability in the promotion of both safety and security.

9.4   Conclusion

It is possible, based on the argument above, that applying an overlapping high reliability management framework to both safety and security missions cannot only enhance both but will also actually protect *against their undermining one another*. The error sensitivity at the foundation of high reliability management can also apply to the identification of and reaction to leading precursor indicators of one mission undermining the other. It is important to think carefully about this enhanced reliability and how it might be lost to both safety and security objectives if these objectives were to be treated as separate processes and managed in separate management domains alone.

References

1. M.F.H. Abulamddi, A survey of approaches reconciling between safety and security requirements engineering for cyber-physical systems. J. Comput. Commun. **5**, 94–100 (2017)
2. A. Boin, D. Smith, Terrorism and critical infrastructures: implications for public-private crisis management. Publ. Money Manag. **26**(5), 295–304 (2006)
3. S. Coursen, Safety vs. security: understanding the difference may soon save lives. *Linkedin* (2014). https://www.linkedin.com/pulse/20140831152519-11537006-understanding-the-diffe rence-may-soon-save-lives-safety-vs-security
4. P. Dombrowski, C. Demchak, Thinking systemically about security and resilience in an era of cybered conflict, in *Cybersecurity Policies and Strategies for Cyberwarfare Prevention*, ed. by J. Richet (ISI Global, Hershey, PA, 2015), pp. 367–382
5. S.H. Jore, The conceptual and scientific demarcation of security in contrast to safety. Eur. J. Secur. Res. (2017). https://doi.org/10.1007/s41125-017-0021-9
6. T. LaPorte, P. Consolini, Working in practice but not in theory: theoretical challenges of high reliability organizations. Publ. Adm. Res. Theory **1**(1), 19–47 (1991)
7. T. LaPorte, High reliability organizations: unlikely, demanding and at risk. J. Conting. Crisis Manag. **4**(2), 60–71 (1996)
8. K.A. Pettersen, P. Schulman, Drift, adaptation, resilience and reliability: an empirical clarification. Saf. Sci. **117** (2016). https://doi.org/10.1016/j.ssci.2016.03.004
9. K.A. Pettersen, T. Bjørnskau, Organizational contradictions between safety and security–perceived challenges and ways of integrating critical infrastructure protection in civil aviation. Saf. Sci. **71**, 167–177 (2015)
10. E. Roe, P. Schulman, *Reliability and Risk: The Challenge of Managing Interconnected Critical Infrastructures* (Stanford University Press, Stanford, 2016)
11. E. Roe, P. Schulman, *High Reliability Management* (Stanford University Press, Stanford, 2008)
12. P. Schulman, The negotiated order of organizational reliability. Adm. Soc. **25**(3) (1993), 353–372

Chapter 10
The Interface of Safety and Security; The Workplace

George Boustras

Abstract 9/11 had a great impact on the development and occurrence of high publicity security-related incidents. One of the biggest impacts was that to public health, due to an increase in psychosocial issues. Cybersecurity incidents and processes of radicalization (either due to religious, political, or economic reasons) can have a direct result on the workplace as well as at the organizational level, which in turn can affect the worker. The aim of this chapter is to explain the main factors linking safety and security, creating a new area for workplace health and safety, that of the "interface of safety and security".

Keywords Safety in the workplace · Security · Safety culture · Security culture

10.1 Introduction

Many high-impact security-related issues have occurred since the turn of the millennium, both large-scale events such as the Paris, Brussels, Nice, London, and Madrid attacks or with a smaller scale such as various knife attacks in Israel and lone wolves. As a category of events, rare episodes of the early noughties (e.g., 9/11, 7/7 in London, Attocha in Madrid) are becoming increasingly "usual". Episodes of violence related to radicalization, cyberattacks, increased fear of a CBRN (e.g., dirty bomb) attack creates a complex security environment also for workplaces. Radicalization [1], which is an emerging issue for workplace health and safety, illustrates the difficulty of Western societies to explain a mechanism that brings to the surface previously unknown forms of societal unrest. Cybercrime as well, which is a product of the large-scale development of information technology, can result in new and unforeseen interactions between previously unrelated places of work. The reliance of modern societies and occupational environments on digital systems illustrates the

G. Boustras (✉)
CERIDES - Excellence in Innovation and Technology, European University Cyprus, Nicosia, Cyprus
e-mail: G.Boustras@euc.ac.cy

potential impact of such attacks [2]. Both radicalization and cyberattacks have in common that they are driven by the human factor.

Moreover, the combined efforts at Critical Infrastructure Protection is a "test bed" to prove the interconnection between these new forms of security threats. These emerging risks have an impact on infrastructure as well as the occupational environment and the employee [3]. At the same time, safety-related issues have been highly affected by the still ongoing economic slowdown [4] and its byproducts (increased occupational psychosocial issues) [5]. Safety in the occupational environment, safety systems and competent authorities are victims of austerity measures, associated with the financial crisis. Financial uncertainty, increased security-related media hysteria result to yet more psychosocial issues.

As Beck [6] discusses the impact of new, globalized risks for individuals in society, it is obvious from the above that a new set of social conditions (cybersecurity threats, radicalization, economic crisis, etc.) that affect the workplace and the employee have been created. These are both a new set of conditions that bring safety and security tighter together and create a more complex and dynamic environment than before. This is a new narrative for workplace safety, constructing a causal link where security-related episodes impact safety, both at the occupational and societal levels. In other words, a new type of risk should be taken into account. A new type of risk that—by default—inherits a large element of uncertainty. In addition, this uncertainty is inherent due to its dependence on human behavior.

The aim of this chapter is to illustrate the hypothesis that safety and security, although they can be seen from different perspectives, have a common interface in the workplace. This chapter takes an exploratory approach to underline and describe this interface. Due to the fact that the author comes from a safety background, forms and schemes related to safety are used. Existing safety science (stereotypes, metaphors, perceptions, theories, publications) still revolve around the basic narrative that stems from the industrial revolution that started in Western societies. Safety focuses on hazards, whereas security focuses on threats. Yet, more new risks and threats and more complex risks and threats are introduced. In this context, it becomes obvious that safety is becoming increasingly dependent on security and is affected by it. *A new scientific domain emerges in the interface between security and safety.*

10.2   Changes to the Physical Environment of Work

Security incidents have dramatic short- and long-term effects on the workplace. Physical injuries and life loss have a direct and immediate impact on day-to-day operations of the organization. Psychosocial issues can have a short, as well as a long-term impact on the organization. Post-traumatic Stress Disorder (PTSD), depression, and other stress-related diseases affect the workplace in an organizational manner as well as leading to financial implications. Stress-related diseases in the workplace have a direct cost to the social insurance system.

It is reported that, further to the collapse of the Twin Towers in New York (2001), a toxic cloud—made of various particles, among them asbestos—covered the wider vicinity [7]. Effects on the over 40,000 first responders and relief workers included short and long-term effects; it has been reported that at least four first responders' deaths are linked to upper respiratory problems, and hundreds of fire-fighters have retired due to health effects [7].

At an organizational level, the effects are significant. Security incidents can be large scale (e.g., 9/11, 7/7, etc.) or small scale (possible security breach in the organization's premises). Risk assessment is reformed to include security aspects as well, especially those linked directly to safety issues. An example of that is the (possible) development of a security policy and/or the establishment of a security recording mechanism (manual or automatic, manned or electronic) in order to avoid (or in response to) an attempted arson attack. Low-risk firms (e.g., office environments) increasingly attempt to consolidate the duties of security and safety officers.

As discussed above, a number of important terrorist attacks have occurred in the last few years such as attacks on the World Trade Center, the London transport system, Paris, Brussels airport and transport system. In addition to the effects described above, attacks on the workplace have further effects. While a security incident is a malevolent—usually preplanned—action, studies have reported that security incidents in the workplace have an impact on the way employees perceive trust toward their employer. The main achievement of the employee, that of the responsibility of the owner/manager to provide a safe and healthy workplace is put in doubt. It is crucial, therefore, to highlight the importance of the integration of security in organizational health and safety planning. The role of the leadership before, during, and after a crisis is crucial for an organization to respond, as well as be in a position to come back to a functional state.

Terrorists tend to choose emblematic employers and workplaces in order to achieve media coverage, also for ideological reasons. The literature suggests that personal preparedness is still at a low level, even in countries that have already faced emblematic security incidents and terrorist attacks [8]. Results suggest that less than half of the population have followed preparatory actions [9]. Security incidents in the workplace tend to create immediate issues with long-lasting effects, yet at the same time, society creates a "mechanism" where these events are eliminated not long after their occurrence. In other words, the emergence of the need for the establishment of a "security culture" similar to the alleged establishment of a "safety culture" at an individual, organizational, and national level becomes important. Based on this argument, one might argue that both security and safety cultures meet as they are targeted firstly to the individual. This is supported by Guldenmund [10] "most empirical studies of safety culture have focused on individual attitudes, perceptions and patterns of behaviours with regard to safety", Mearns and Yule [11] "Having safety as a central value would be the defining moment for any organization embarking on the development of a positive safety culture, irrespective of the national context it is working in", and the analysis of Smith [12] describing the switch of security (after the end of the Cold War) from the national level to the community and individual.

10.3   Cyber-Related Issues in Relation to Safety
in the Workplace

Cyberattacks occur more and more often. Cyberattacks can have a significant impact on the workplace and the welfare of the employee. A possible attack on the workplace may mean loss of sensitive data that may lead to a temporary or permanent closure of the firm. High-risk operations may be affected by cyberattacks with the possibility of operational or physical damage. News titles after the 2016 Brussels attacks reported preparatory work for a possible terrorist operation in one (or more) of the nuclear installations in Belgium.

Cyberattacks can also cause serious damage at the societal level. Particular emphasis should be paid on the relationship between cybersecurity in the workplace and the safety and reliability of Critical Infrastructures. Energy Critical Infrastructure is the main supporting pillar of National and International economic activity. Possible interruptions can cause serious damage to the wider workplace [13]. Domino effects from possible interruptions to Critical Infrastructure can cause serious financial damage, among others. "Hundreds of shops across south-east London and north Kent were forced to close and commuters spoke of 'incredibly frightening' conditions on the roads as traffic lights failed" [14]. Modern infrastructures operate as a "system of systems" with many interactions, interconnections, and interdependencies among these systems. Thus, damage occurred due to cyberattacks in the workplace of one infrastructure system can cascade and result in failures and cascading effects onto all related and dependent infrastructures eventually impacting the broader economy and society [15]. Such interconnections and interdependencies can be digital, physical, geographical, or logical [16].

10.4   Radicalization Effects on Safety in the Workplace

An emerging risk about workplace safety and security is that of radicalization. Radicalization is a process where individuals are subjected to extremist material with a direct effect on their social behavior, perceptions about society and justice. Various mechanisms such as personal grievance/revenge, existing or developing psychological issues explain the emergence of radicalized individuals to terrorists [17]. Despite media reports that the perpetrators of the majority of the latest terrorist incidents were "homegrown, radicalized youth", there is no literature highlighting the relationship between radicalization and the occupational environment. This chapter will attempt to underline the various parameters affecting everyday life in the workplace, in relation to radicalization; it will attempt to explain the common factors leading to the occurrence of occupational safety issues. The relationship between human behavior and safety in the workplace is straightforward [18]. Rational or irrational behaviors lead to accident occurrence. Human behavior in turn comprises a number of attributes.

Reports from captured terrorists mention various reasons that (allegedly) led them to become radicalized. Among the causal factors mentioned are various forms of discrimination (mainly dealing with religious or political issues) and bullying (because of the discrimination). Discrimination in the workplace has a long history associated with it. Huang and Kleiner [19] mention that *"In the 1960s and 1970s, blacks and women fought for their rights. In the 1980s and 1990s, it was gays and lesbians. Now it has turned into employers and employees and the battlefield is religion in the workplace"*. Religious discrimination [20] in the workplace can have a formal and informal way of occurrence. Religious jokes, exclusion due to religious issues, discounting of religious beliefs can create an environment where personal grievance and willingness to take revenge are cultured. Irrespective of the intensity of the comments or behaviors in the workplace, perceived discrimination [21] can have various effects. Whether discrimination is wide or covert, what matters most is how the employee will perceive it. The results of perceived exclusion can be felt in the workplace and the society. Exclusion and discrimination (as it will be analyzed in another part of this chapter) is a common factor leading to (among others) workplace issues. Verkuyten [22] emphasizes the role of discrimination as a leading factor to radicalization. Popular press describes terrorists as "normal" people until a certain age (prior to their actions) with informal, precarious, or formal types of employment. Bullying/discrimination due to (religious) discrimination has dual effects to the workplace (absenteeism, with obvious social and economic impacts) as well as to the victim (psychosocial issues).

10.5   Financial Crisis Influence on the Increase of Psychosocial Issues

The economic crisis that started in 2008 with the collapse of Lehman Brothers left an impact on specific parts of the Western World. Southern European countries suffered the majority of the impacts of the crisis [4]. For some countries, the economic crisis became a social crisis as well.

This chapter argues that the effects of the financial crisis can be a connecting factor between safety and security. Undoubtedly, the financial crisis has had an impact on the job market [23] and public health [24]. Changes in the job market meant an increase in unemployment, temporary and undeclared work. Lower wages, insecurity, informal working hours paint a dramatic picture with clear impacts on health and safety of employees. A direct impact is that of the increase of psychosocial issues in the workplace [25, 26]. Psychosocial issues in the workplace have direct effects on the organization and the workforce. Anxiety, bullying, mobbing, depression lead—among others—to an increase in occupational injuries. Insecurity, desperation lead to various forms of extreme behaviors, creating the background for potential attacks and aggression, which in turn leads to social exclusion. At this point, it should be noted that there are similarities with the process leading to radicalization, as described above. Historical analysis shows that financial crises cause political disruption and political radicalization [27]. The economic crisis has been the driving force behind the rise of extremist political parties in Europe, which marketed anti-globalization, anti-immigration, anti-foreigner rhetorics [28]. *Social exclusion surfaces as the connecting factor between potentials leading to safety- and security-related incidents.*

10.6   Conclusions

The aim of this chapter was to describe factors that are redefining safety in the workplace; the interface of safety and security. The chapter has presented arguments that showed how specific factors and cases (cybersecurity, critical infrastructure protection, radicalization, etc.) can be the "connecting dots" between security and safety. Still, without the empirical analysis necessary to establish causal connections, the factors explored indicate how the two (safety and security) increasingly meet at the level of the individual worker, with both causal connections to and ramifications for the organization. In this perspective, many workplace risks must be viewed as highly dependent on human behavior and an outcome of various psychological processes (e.g., developmental, cognitive). For example, psychosocial issues relate to security incidents (and relevant media hysteria), agitating workers and creating prolonged feelings of fear and anxiety. Also, as discussed in the chapter, political radicalization and the financial crisis have been shown to have social exclusion as a common route that can lead to security incidents. Consequently, we must consider how to approach the participation of workers in safety and security, as workers previously are viewed as part of safety efforts. Among other, formal ways of worker participation in the development of safety policies, through the legally binding establishment of health and safety committees, could be replicated toward the establishment of a security culture in the workplace. Training, risk assessment, adoption of policies should be typical for both safety and security cultures. However, a fundamental difference between the two (safety and security) lies in the fact that with safety, there is a legal requirement to the owner/manager—in other words, responsibility is personalized. With security, this is not the case as the State authorities provide the backbone. This again has consequences for strategic priorities and the development of incentives.

References

1. S.A. Høyland, Exploring and modeling the societal safety and societal security concepts–a systematic review, empirical study and key implications. Saf. Sci. (2017). ISSN 0925-7535. https://doi.org/10.1016/j.ssci.2017.10.019
2. B. Wilpert, Impact of globalization on human work. Saf. Sci. **47**(6) (2009), 727–732. ISSN 0925-7535. https://doi.org/10.1016/j.ssci.2008.01.014

3. M. Nilsen, E. Albrechtsen, O.M. Nyheim, Changes in Norway's societal safety and security measures following the 2011 Oslo terror attacks. Saf. Sci. (2017). ISSN 0925-7535. https://doi.org/10.1016/j.ssci.2017.06.014

4. I. Anyfantis, G. Boustras, A. Karageorgiou, Maintaining occupational safety and health levels during the financial crisis–a conceptual model. Saf. Sci. (2016). ISSN 0925-7535. https://doi.org/10.1016/j.ssci.2016.02.014

5. H.H.K. Sønderstrup-Andersen, E. Bach, Managing preventive occupational health and safety activities in Danish enterprises during a period of financial crisis. Saf. Sci. (2017). ISSN 0925-7535. https://doi.org/10.1016/j.ssci.2017.03.022

6. U. Beck, *Risk Society, Towards a New Modernity* (Sage Publications, 1992)

7. EHS Today, 9/11: Safety and Health Lessons Learned (2006). http://www.ehstoday.com/fire_emergencyresponse/ehs_imp_38472. Visited on 28 February 2018

8. M. Kano, M.M. Wood et al., Terrorism preparedness and exposure reduction since 9/11: the status of public readiness in the United States. J. Homeland Secur. Emerg. Manage. 8(1) (2011)

9. J.L. Gin, J.A. Stein, K.C. Heslin, A. Dobalian, Responding to risk: awareness and action after the September 11, 2001 terrorist attacks. Saf. Sci. **65**, 86–92 (2014)

10. F.W. Guldenmund, The nature of safety culture: a review of theory and research. Saf. Sci. **34**(1), 215–257 (2000)

11. K. Mearns, S. Yule, The role of national culture in determining safety performance: challenges for the global oil and gas industry. Saf. Sci. **47**, 777–785 (2009)

12. S. Smith, The increasing insecurity of security studies: conceptualizing security in the last twenty years. Contemp. Secur. Policy **20**(3), 72–101 (2007). https://doi.org/10.1080/13523269908404231

13. C. Varianou Mikellidou, L.M. Shakou, G. Boustras, C. Dimopoulos, Energy critical infrastructures at risk from climate change: a state of the art review. Saf. Sci. (2017). ISSN 0925-7535. https://doi.org/10.1016/j.ssci.2017.12.022

14. P. Dominiczack, A. Nassif, 60,000 homes are blacked out as vandal fire causes huge power cut, Evening Standard (2009). https://www.standard.co.uk/news/60000-homes-are-blacked-out-as-vandal-fire-causes-huge-power-cut-6746359.html. Last visited 19 March 2018

15. European Climate Adaptation Platform, Adaptation information, January 2018. http://climate-adapt.eea.europa.eu/adaptation-information/general

16. S.M. Rinaldi et al., Identifying, understanding, and analyzing critical infrastructure interdependencies. IEEE Control Syst. **21**(6) (2001)

17. C. McCauley, S. Moskalenko, Mechanisms of political radicalization: pathways toward terrorism. Terror. Polit. Violence **20**(3), 415–433 (2008). https://doi.org/10.1080/09546550802073367

18. T.B. Sheridan, Forty-five years of Human–Machine systems: history and trends. Keynote Address, in *Proceedings of 2nd IFAC Conference on Analysis, Design and Evaluation of Human–Machine Systems. Varese, Italy, September 10–12, 1985* (Pergamon, Oxford, 1986)

19. C.C. Huang, B.H. Kleiner, New developments concerning religious discrimination in the workplace. Int. J. Sociol. Soc. Policy **21**(8/9/10) (2001), 128–136. https://doi.org/10.1108/01443330110789880

20. B.R.E. Wright, M. Wallace, J. Bailey, A. Hyde, Religious affiliation and hiring discrimination in New England: a field experiment. Res. Soc. Stratif. Mob. **34** (2013), 111–126. ISSN 0276-5624. https://doi.org/10.1016/j.rssm.2013.10.002

21. J.I. Sanchez, P. Brock, Outcomes of perceived discrimination among Hispanic employees: is diversity management a luxury or a necessity? Acad. Manag. J. **39**(3), 704–719 (1996). https://doi.org/10.2307/256660

22. M. Verkuyten, Religious fundamentalism and radicalization among Muslim minority youth in Europe. Eur. Psychol. **23**, 21–31 (2018). https://doi.org/10.1027/1016-9040/a000314

23. N. Drydakis, The effect of unemployment on self-reported health and mental health in Greece from 2008 to 2013: a longitudinal study before and during the financial crisis. Soc. Sci. Med. **128** (2015), 43–51. ISSN 0277-9536. https://doi.org/10.1016/j.socscimed.2014.12.025

24. M. Karanikolos, P. Mladovsky, J. Cylus, S. Thomson, S. Basu, D. Stuckler, J.P. Mackenbach, M. McKee, Financial crisis, austerity, and health in Europe. Lancet **381**(9874), 1323–1331 (2013). https://doi.org/10.1016/S0140-6736(13)60102-6
25. N. Mucci, G. Giorgi, M. Roncaioli, J. Fiz Perez, G. Arcangeli, The correlation between stress and economic crisis: a systematic review. Neuropsychiatr. Dis. Treat. **12**, 983–993 (2016). https://doi.org/10.2147/NDT.S98525
26. V. Sedano de la Fuente, M.A. Camino López, I. Fontaneda González, O.J. González Alcántara, D.O. Ritzel, The impact of the economic crisis on occupational injuries. J. Saf. Res. **48** (2014), 77–85. ISSN 0022-4375. https://doi.org/10.1016/j.jsr.2013.12.007
27. M. Funke, M. Schularick, C. Trebesch, Going to extremes: politics after financial crises, 1870–2014. Eur. Econ. Rev. **88** (2016), 227–260. ISSN 0014-2921. https://doi.org/10.1016/j.euroecorev.2016.03.006
28. G. Calhoun, G. Derluguian, *The deepening crisis: governance challenges after neoliberalism* (New York University Press, New York and London, 2011)

Chapter 11
Exploring the Interrelations Between Safety and Security: Research and Management Challenges

Corinne Bieder and Kenneth Pettersen Gould

Abstract This chapter discusses some of the research and management challenges related to the safety and security nexus. In the first part, we address the conceptual connections between safety and security and discuss how different perspectives on how they come together allows for characterizing the complexity and ambivalence of their interrelations. We then go on to identify tradeoffs between safety and security and show that these exist both in theory and practice. Managing both safety and security means tradeoffs and power relations between internal entities and professionals, but also beyond its own boundaries since some vulnerabilities escape the organization's scope. In the final part of the chapter, we argue that addressing the interrelations between safety and security poses managerial and research challenges that call for global approaches to apprehend the multiple facets of the issue. We explain that little has been done on how the global trends of the risk society bring with them unanticipated and "hidden" effects on organizations safety and security practices and that it is here, as a macro-global oriented approach to organizational safety and security research, that the two fields of safety and security confront a shared research agenda.

Keywords Safety · Security · Management · Risk · Societal safety · Societal security

11.1 Introduction

Following recent events and disasters, and as threats and hazards are defined more and more as systemic risks and products of modern society, safety and security are coming together both in regulation and the ambitions of management. What does

C. Bieder (✉)
ENAC (French Civil Aviation University), University of Toulouse, Toulouse, France
e-mail: corinne.bieder@enac.fr

K. Pettersen Gould
University of Stavanger, Stavanger, Norway
e-mail: kenneth.a.pettersen@uis.no

© The Author(s) 2020
C. Bieder and K. Pettersen Gould (eds.), *The Coupling of Safety and Security*, SpringerBriefs in Safety Management,
https://doi.org/10.1007/978-3-030-47229-0_11

it mean that safety and security become mixed in the management of hazardous technologies and activities? Despite their apparent or intuitive proximity when considered conceptually, safety and security reveal some nuances and differences when leadership and practices are analyzed [1]. Technologies and activities also differ in the number and types of threats and hazards they have to deal with. Answers to the question are thus more complex than the initial conceptual similarities indicate. Looking at synergies and tensions between the two, in this final chapter, we come back to some main insights derived from the previous chapters and discuss challenges of addressing such a multi-faceted issue from both research and management viewpoints.

11.2   Implications of Definitions

From both a research perspective and an angle of management, defining safety and security seems a natural place to start in order to address their interrelations. Many seem to expect a unified understanding of definitions. This could allow both different scientists and practitioners across sectors to "speak the same language", in order to better understand each other and work together. However, merely based on the various outlooks represented in this book, conceptual agreement is easier said than done. Brooks and Cole (this volume) also identify clear distinctions in the underlying body of knowledge between the safety and security professions, even though some overlap exists around the management of risks. Perhaps shared definition is not a sensible goal to pursue at all, as safety and security knowledge vary to a large degree depending on, among other, hazards/threats, disciplinary approach, regulatory context, and practice.

11.2.1   Defined as What We Want to Prevent

Both safety and security are seen as the freedom from harm. Indeed, when successful as management strategies, both safety and security lead to prevention or minimized unwanted consequences for people, the environment, and/or property. In other words, both share a common goal in loss prevention. However, when focusing on unwanted events and the associated causal factors that are eliminated and constrained, safety and security differ. In all the chapters, it is acknowledged that the involvement of human intent as a cause marks an important difference in how events are considered, managed, and prevented, justifying a clear distinction between safety and security. Although here again, there are some nuances. Where Blokland & Reniers, Leveson, Bongiovani, and Boustras (this volume) put intentionality forward as the main distinction, Jore, Brooks & Cole, Wipf, and LaPorte (this volume) are more nuanced, arguing that the difference is most clearly defined by malicious intent or the "primacy of hostile intent" (Schulman this volume) as a criterion for a security event.

11.2.2 Risk as an Overarching Framework for Management

According to Leveson and Bongiovanni (both this volume), conceptual exercises around the terms safety and security are not what really matters. In Leveson's approach (this volume), safety and security are both considered as sources of a system's loss of control, with no need for any distinction in the way they are considered and managed. Security events have an equivalent status to that of technical failure or human error. In this approach, the theoretical, methodological but also practical consequences of considering both safety and security are minimized. The two aspects smoothly combine by adding security threats to safety hazards and the associated failure scenarios to that already identified in a safety risk analysis. Eventually, from system engineering and design thinking perspectives, or more specifically, from the perspective of how to design controls in a system, addressing safety and security together and in the same way, seems natural and achievable. However, defining both safety and security as a dynamic control problem is a way of framing the issue of their interrelations that makes the solution already available and implemented for safety applicable to security as well. As stated during the workshop, we must be careful of "defining the problems based on our solutions".

Blokland and Reniers (this volume) apply a broadly used definition of risk as a reference to define safety and security. A key challenge in the movement from safety/security to risk is that both safety and security may improperly communicate an absolute degree of freedom from risk that is not implicit in risk science [2]. Many of the biggest safety and security risks we currently face are, although different in their causes, all products of human activities seen as being necessary. Thus, risks emerge from activities that we have to or want to undertake and cannot be "managed away" by science. Political demands for both safety and security may confuse both the politicians themselves and regulators, creating unrealistic and unwarranted expectations for action. The margins of error may be changed but the risk will remain as long as the activities continue. This has influenced and broadened approaches to uncertainty in risk research [3]. A significant issue for the assessment and management of risks is whether uncertainty is positive or negative. For example, in terms of nuclear power plants or airlines, is it good or bad to have uncertainties? Is it different depending on threats and hazards? A growing body of work has acknowledged that uncertainty plays a significant role in our understandings of safe and secure systems and societies [4–6]. In addition, how much an activity means for us is strongly associated with our judgment as to whether the risk is worth taking [7]. Consequently, both the risk appraisal and risk management of threats and hazards are influenced by individual and social factors. For situations where there is a rise in demand for safety and security, such as after a major disaster or a terrorist attack, we should expect the perception of risks to be amplified and standards to become more stringent than compared to direct estimation. However, for situations where demands for safety and security are low, such as for activities that produce large short-term benefits where consequences are uncertain, for example, mobile digital communication technologies, we can expect risks to be attenuated and standards to be more relaxed [2, 8].

11.2.3 In Safety We Trust, in Security We Distrust

Security management, whether through its interrelations with safety or not, induces some organizational challenges from both a research and a practical viewpoint. Whereas safety, at least in principle, has reached a state where openness and sharing of information are acknowledged as criteria of improvement, security is rather a world of secrecy, both for attackers and for the potentially targeted organizations, which avoids increasing their vulnerability by unveiling characteristics. One of the categories of security threats that have received increased attention is insider events, as sadly illustrated by the Germanwings catastrophe in 2015 where the crash was caused deliberately by the co-pilot. The possibility of having employees that are on the one side trusted for their contribution to safety and at the same time distrusted as potential security threats represents a challenge for organizations [1], both conceptually and in practice as advanced by Jore (this volume) on the relevance of how to develop a security culture in organizations. Moreover, gaining access to the field for researchers, or even exchanging information with experts, becomes a challenge as soon as security is involved.

Consequently, addressing the interrelations between safety and security is not as simple as "mixing" the two using a broad risk approach or extending the scope of existing either safety or security approaches. As illustrated by Schulman (this volume), security involves vulnerability variables that are outside the boundaries of organizations. This observation has several implications: practical and managerial ones within an organization, but also methodological when it comes to describing, analyzing, and understanding the interrelations between organizations and their increasingly global environment. Addressing the interrelation between safety and security thus requires some caution not to be blinded by conceptual elegance or methodological solutions already available.

11.3 Tradeoffs Between Safety and Security

Exploring the synergies and tensions between safety and security allowed for recognizing the respective professions associated with safety and security, the resources that they both strive for, and lastly some conceptual and methodological rivalry between scientific frameworks and communities. These issues highlight the point that although organizations need a degree of both safety and security, there may be some important tradeoffs between them.

The security world, in terms of the industries, the actors, and the tasks involved, has significantly changed. "It used to be a police concern, classified information (…). Today, security is something a lot of people are doing (private companies, civil society…). It encompasses the security guard at the hotel, at the airport" (Jore, this volume) as well as others like the corporate security officers in companies. The spreading of security as an activity has been accompanied by the structuring of security as a profession (Jore, this volume), with tasks, underlying body of knowledge, and professional communities distinct from safety (Brooks & Cole, this volume). With security becoming an increased concern for hazardous industries, safety and security are now more than ever competing within organizations in terms of attention, resources, and power. As advanced by LaPorte (this volume), the form these managerial challenges can take depends on the historical path of the organization with respect to safety and security. Particularly, challenges are formed by which one existed first in the organization and what mix is already established. Introducing a new management function, whether safety or security, involves allocating dedicated resources that may also change the existing power balance and share of voice between the respective functions within organizations.

Beyond these areas of potential tension, the apparent proximity between safety and security can make it tempting for researchers to extend the scope of concepts and methods from one area of expertise (whether safety or security) to the other. As an illustration, one can take the evolution of the European Commission's research agenda for aviation. Safety was historically the core topic of the Advisory Council for Aviation Research and innovation in Europe (ACARE). Over the past few years, the safety working group has become the Safety and Security working group, and the 2017 version of the Strategic Research and Innovation Agenda[1] deriving from the ACARE vision includes a large security section with a research budget emphasizing security. Although safety and security are close enough to appear in a shared research agenda, the intersection between safety and security is still largely addressed by experts in one or another domain, using concepts, theories, and methods coming from either safety or security. What we may be forgetting, ironically enough, is that doing so leads to transferring not only research approaches but also their (tacit) foundational premises that often remain implicit. For example, safety strives for everybody within the organization sharing information. Also, safety research is based on many well-developed researcher–practitioner collaborations. Looked through a security lens, the same premises do not seem to be considered foundational. Consequently, although transferring and adapting safety approaches to security may be tempting, understanding the underlying premises may be essential for this work to be successful. Of course, a parallel rationale applies for the opposite move from security to safety.

---

[1] https://www.acare4europe.org/sites/acare4europe.org/files/attachment/acare-strategic-research-innovation-volume-1-v2.7-interactive-fin_0.pdf.

11.4 The Societal Convergence of Safety and Security

Returning to the chapters of the book, this volume shows that the management of safety and security in organizations is not isolated from social expectations and societal changes. As stated by Brooks and Coole (this volume, p. 64), "society becomes more complex and its members more risk averse", whether the risk is associated with safety or security. Whereas the level of safety and security has increased over time, perceptions of risk have evolved differently [9], and overall acceptability of risk has decreased. The safest assumptions for the future seem to be that technology but also society will increase in complexity, uncertainty will abound, and new vulnerabilities will emerge. We believe that safety and security in organizations are not disconnected from the wider patterns of neo-liberal influence [10] characterized by extensive deregulation, privatization, and outsourcing. Today individuals have a lot more objectives and a lot more to lose than in the past, while at the same time their trust in institutions has decreased. These developments of the risk society are quite well described within European sociology [11] and in risk research [12]. Also, there are some issues that we can associate with the risk society thesis that have been picked up by safety and security research. One such issue is the acknowledgment that it is impossible to anticipate and control everything *a priori* and that other (complementary) strategies are needed. This line of thought recognized and addressed by HRO theory [13] in the late 80s and early 90s, the mindful organizing developments [14] and the resilience engineering research strand more recently [15], challenge management models based on logics of hierarchical control and converge on shared issues such as: How to be prepared to be surprised? How to manage the unexpected? What would it take to acknowledge uncertainty and evolve accordingly at a societal level, but also at the various organizational levels? However, as several scholars have already pointed out in relation to safety [10, 16], little has been done on how the global trends of the risk society bring with them unanticipated and "hidden" effects on organizations' safety and security practices. It is perhaps here, as a macro-global approach to organizational safety and security research, that the two fields more fully confront a shared research agenda and to which several of the authors in this volume propose key contributions. For example, LaPorte (this volume) provides an interesting angle in the obligations of leadership to reach beyond the boundaries of their own organizations to avoid the "thinning of watchfulness", "assure organizational and public understanding of safety and security stewardship roles and their fundamental contributions" as well as to "enhance a sense of honor and resources—beyond operational costs—for safety and security stewards" (p. 84). The theoretical developments in safety science during the 90s [17–19] underlined the impact of organizational and institutional aspects, with timeframes that went significantly back in time and upstream from operations. With security, threats such as terrorist attacks may for both explanatory and protective purposes have to be related to even longer-developing phenomena, such as radicalization as advanced by Boustras (this volume), with timeframes that go beyond technological design, organizational decisions, regulations, and laws.

11.5   Developing a Global Approach to Organizational Safety and Security Research

As indicated by several of the considerations in this volume (Schulman, Boustras), if we are to develop a global approach to study organizations, we need to consider different scales at the same time. Even more so, the range of scales needed to understand the interactions between safety and security may be even greater than when addressing safety or security alone. Pettersen and Bjørnskau [1] pointed at the impact of EU security regulations on aviation organization employees' working conditions and influence on safety in the field. Schulman (this volume) underlines the limitations of focusing exclusively on organizations when addressing security, for these organizations have little if any control on external vulnerability variables. Also, Boustras (this volume) discusses the need for multiple scales related to the link between radicalization as an international issue and workplace risks at an individual level. Conversely, with ever more interconnected systems and critical infrastructures, an individual malevolent act can have significant worldwide safety consequences, either direct, indirect, or both. These interactions between phenomena and actors at very different levels raise questions. For example, what is the motivation (even before the question of the means) for a company to prioritize and dedicate its own resources to preventing a malicious attack that may be high on the agenda as societal issues but with a very low probability for incidents in a specific company? Also, in the example of the Brussels airport attack in 2016, for example, the Head of Brussels airport security reported that members of his crisis team had a hard time getting to the crisis room from outside the airport for security at the site was under government responsibility. The scope of responsibility and leeway to manage safety, security, and their interrelations during a crisis is thus another issue.

In addition to the challenge of multiple scales, another issue is the need to address a multitude of actors at the same time, as well as being clear on what actors and at which levels research will be conducted. Taking a macro approach also suggests the inclusion of actors that are not so commonly considered in safety and security research. For example, Brooks and Cole (this volume), by addressing synergies and tensions through the lenses of the associated safety and security professions, suggest that both these distinct professional communities need to be considered as such. Likewise, as illustrated by Wipf (this volume), the attacker, that doesn't "exist" as such from a safety viewpoint, becomes an actor to be considered. Another actor suggested by Bongiovanni (this volume) to be considered is the end-user. Although central to the design thinking approach through which he looks at safety–security convergence, the passengers are largely absent from safety management approaches within aviation.

As a final remark, relevant empirical descriptions of how global trends bring with them unanticipated and "hidden" effects on organizations' safety and security practices are still few. Part of the reasons may be the research environment that needs to be created to have access to security-related aspects, or the tradeoffs between safety and security research topics and communities. Others may be, as discussed above, related to the challenges of addressing multiple scales, dimensions, and aspects. In order to minimize "biases", empirical research would best be framed and conducted by research teams involving both safety and security researchers. Bringing them together will (hopefully) allow for better definitions of scales, timeframes, and actors that are relevant for revealing how global trends bring with them unanticipated and "hidden" effects on organizations' safety and security.

References

1. K.A. Pettersen, T. Bjornskau, Organizational contradictions between safety and security—Perceived challenges and ways of integrating critical infrastructure protection in civil aviation. Saf. Sci. **71**, 167–177 (2015)
2. J.F. Short, *Organizations, Uncertainties, and Risk* (Westview Pr, 1992)
3. K.A. Pettersen, Understanding uncertainty: thinking through in relation to high-risk technologies, in *Routledge Handbook of Risk Studies*, ed. by A. Burgess, A. Alemanno, J.O. Zinn (Routledge, London, 2016)
4. G. Grote, Management of Uncertainty: Theory and Application in the Design of Systems and Organizations (Springer Science & Business Media, 2009)
5. N.N. Taleb, *The Black Swan: the Impact of the Highly Improbable* (Random House, New York, 2007)
6. T. Aven, *Risk, Surprises and Black Swans: Fundamental Ideas and Concepts in Risk Assessment and Risk Management.* (Routledge, 2014)
7. B. Ale, *Risk: An Introduction: The Concepts of Risk, Danger and Chance* (Routledge, 2009)
8. R.E. Kasperson, O. Renn, P. Slovic, H.S. Brown, J. Emel, R. Goble, J.X. Kasperson, S. Ratick, The social amplification of risk: a conceptual framework. Risk Anal. **8**(2), 177–187 (1988)
9. N. Pidgeon, R. Kasperson, P. Slovic, *The Social Amplification of Risk* (Cambridge University Press, 2003)
10. N. Pidgeon, Observing the English weather: a personal journey from safety I to IV, in ed. by J.C. Le Coze, *Safety Science Research: Evolution, Challenges and New Directions*, (CRC Press, 2019), pp 269–280
11. U. Beck, *Risk Society: Towards a New Modernity* (Sage, 1992)
12. K.L. Henwood, N. Pidgeon, *Risk and Identity Futures. Future Identities Programme* (Government Office of Science, London, 2014)
13. T.R. LaPorte, P. Consolini, Working in practice but not in theory: theoretical challenges of high-reliability organizations. J. Public Adm. Res. Theor. **1**, 19–47 (1991)
14. K.E. Weick, K.M. Sutcliffe, *Managing the Unexpected*, vol. 9 (Jossey-Bass, San Francisco, 2001)
15. E. Hollnagel, D.D. Woods, N. Leveson, (eds.), *Resilience Engineering: Concepts and Precepts* (Ashgate Publishing, Ltd., 2006)
16. J.C. Le Coze, Globalization and high-risk systems. Polic. Pract. Health Saf. **15**(1), 57–81 (2017)
17. T.R. LaPorte, High reliability organizations: unlikely, demanding and at risk. J. Contingencies Crisis Manag. **4**(2), 60–71 (1996)
18. J. Rasmussen, Risk management in a dynamic society: a modelling problem. Saf. Sci. **27**(2), 183–213 (1997)
19. J. Reason, *Managing the Risks of Organizational Accidents* (Ashgate, Aldershot, 1997)

Printed in the United States
By Bookmasters